Published by The Educational Publisher, Inc.
1313 Chesapeake Ave
Columbus, OH 43212
United States of America
www.BiblioPublishing.com
www.EduPublisher.com
ISBN: 978-1-62249-084-4

World War Two, Cause and Effect contains much to like, especially for those history buffs who thrive on detail. Mr. Brady's coverage of 'the greatest war in world history' includes coverage of land, sea and air forces, battles, material and troops. It also explores intricate details of the 'why' of WW2, its genesis in the treaty of Versailles and humiliation of the German people, and how Adolf Hitler took full advantage of that German feeling of marginalization.

The book tracks such arcana as the scuttling of the Graf Spee and its implications, Hitler's seeming passive-aggression in not destroying British troops at Dunkirk, the armament of various tanks and battleships and their vulnerability to shells and even the implications of commanders' decisions to attack civilians instead of fuel supplies.

The book avoids political cant, though it does show some long-term implications of certain political decisions by Truman, Churchill and others. Truman in using the 'A' bomb in opposition to Eisenhower who advised against it, for example. Churchill decided to help keep Russians fighting, thus diverting Germans from attacking England.

The weakness of the book may be its strength, that details take the place of sweeping narrative. Again, if military history is a reader's desire, those details will satisfy. This reader took issue with only a few items: the speculation about Vietnam and America's hesitation to cut its losses there in light of Pearl Harbor is a stretch. On page 89 the author might have used 'day of infamy,' instead of 'fateful day in history,' though either works. There are a few repeated phrases here and there, and the final two chapters seem to be an afterthought. All told a richly detailed, almost scholarly treatise on World War Two and its aftermath.

Byron Edgington
Author of *The Sky Behind Me, a Memoir of Flying and Life.*

About the Author

Bill Brady comes from a professional Engineering background and worked in senior management positions for companies such as Coca-Cola and General Electric. Following a successful career in these international companies, he subsequently established his own management consultancy, where he was engaged in various projects and assignments, not only in South Africa, but also in Zambia and Nigeria.

Bill then retired, but came out of retirement to join another international company and presently conducts specialised training in Operations Management and Finance at MBA level.

Bill always had a lifelong passion for history and having majored in that discipline, he developed a passion for the history of World War Two, in particular. Becoming a member of The South African Military History Society of Kwa Zulu Natal based in Durban. Bill rose to the position of Chairman; a position he still holds.

Bill presents papers on military history to professional bodies and academia on a regular basis. This compendium is a concatenation of the papers presented and as such is a valuable collection of many of the events, actions, battles and biographies of personalities of World War Two, making it an easy matter for both the serious and casual student of history to obtain an understanding of that particular aspect of history, without having to read a tome.

They shall grow not old, as we that are left grow old:
Age shall not weary them, nor the years condemn.
At the going down of the sun and in the morning,
We will remember them.

This book is dedicated in gratitude to my wife Kathleen for her tolerance.

We have been blessed with two lovely daughters, Lorraine and Michelle who have always been a source of great pride and joy.

We thank them for giving us wonderful grandchildren in Devin, Tayla, Chloe and Ava.

CONTENTS

Overview of World War Two from a South African Perspective

Hitler's rise to power in 1933 saw the repudiation of the Versailles Treaty, the re armament of Germany and the road to war. Nazi Germany invaded Poland in 1939 to launch the Second World War.

When British Prime Minister Neville Chamberlain announced on Sunday September 3rd 1939 that Britain was at war with Germany, South Africa was once more divided. Prime Minister Hertzog opted for neutrality as he had done in 1914.

The cabinet was split; Deputy Prime Minister Smuts proposed that South Africa sever relations with Hitler's Third Reich and join the Allies. Hertzog's neutrality motion was defeated in Parliament by thirteen votes, Hertzog resigned. Governor General Sir Patrick Duncan invited Smuts to form a new government. On September 6th South Africa declared war on Nazi Germany.

1940: was a year of terrible disasters for the Allies.
The Germans invaded Denmark and Norway in April; the Anglo-French response was nothing short of a fiasco.

An even greater disaster fell in France when the Nazi's invaded the Low Countries and France in May, by-passing the Maginot Line. By this stage Winston Churchill had replaced Chamberlain as Prime Minister and steeled the people to continue the fight.

The "miracle of Dunkirk" saw three hundred thousand troops plucked from the beaches, but they lost all their equipment. Churchill flew to France to urge them not to give up, but the collaborators are in the majority. After Italy entered the war, the French surrendered.

Britain braced herself for invasion. Sea power was no longer the primary issue. It was airpower, and to achieve air superiority the Luftwaffe had to neutralise the RAF.

Many South Africans, including Sailor Malan with thirty five kills came to Britain's rescue during the Battle of Britain. The Germans failed to gain air superiority and the invasion was cancelled. Churchill rallied the people with his famous speech, "We will fight them on the beaches, the landing grounds, in the fields in the hills and in the streets, and we shall never surrender!" The response to this was a rapturous applause in the House of Commons, and reinforced the bulldog spirit of the British people.

More than two million South Africans volunteered for service including one hundred and twenty thousand non whites.

1941: Heralded the intensification of the "Blitz."
Many British cities were flattened and the evacuation of children to South Africa is hastened. The Osewa Brandwag commits many acts of sabotage and attack service men in uniform.

Hitler launches his attack on Russia in June. This proved to be a terrible blunder, the Germans did not prepare for winter warfare.

U – Boats decimate British merchant shipping in the Atlantic; US President Roosevelt approves the Lend-Lease programme to help Britain survive.

General Smuts is promoted to Field Marshall on his 71st birthday; South African forces complete the conquest of the Italians in East Africa.

However, at Sidi Razegh in North Africa the South African 5th Brigade encountered strong enemy forces and were badly mauled by the Afrika Korps with over three thousand troops captured.

Robey Leibrandt the former heavyweight boxing champion returns to South Africa as a German agent with orders to assassinate Prime Minister Smuts. He is betrayed, captured and sentenced to death for High Treason.
Imperial Japan delivers a devastating blow at Pearl Harbour on the 7th December 1941. Over three hundred and fifty aircraft from six Japanese carriers decimate the American Pacific Battle Fleet to inflict the worst military defeat on the United States in its two hundred year history.

As predicted by Smuts; on the 10th of December, just three days after Pearl Harbour, HMS Prince of Wales and HMS Repulse are sunk off Malaya.

At the same time on the other side of the world, the Germans, after massive victories, suffer a severe reverse before the gates of Moscow.

1942: Started With a Series of Disasters for the Allies.
At Singapore on15th February, the greatest ever defeat of British arms occurs when one hundred and thirty thousand British troops surrender to thirty thousand Japanese. The Americans capitulate in the Philippines and the Dutch in Indonesia.

South Africa earned the first of four Victoria Crosses of the Second World War when Squadron Leader John Nettleton led a raid of Lancaster bombers on Germany.

At Tobruk, Major General Klopper, surrenders with thirty thousand men, including ten thousand South Africans.
Durban and Pietermaritzburg are "blacked out" following sightings of Japanese sea planes.

The first thousand bomber raid is mounted by the RAF on Cologne; Colonel Doolittle bombs Tokyo and the US fleet sinks four Japanese aircraft carriers at Midway. South African troops invade Madagascar.

For the South African forces, the great event is at El Alamein where they fought with Montgomery's 8th Army in what Churchill described as the "End of the Beginning".

The second South African VC of the war is awarded to Sergeant Quentin Smythe of the Natal Carbineers for conspicuous valour at Alem Hanza in North Africa on 5th June1942.

In November the Russians launch their counter attack at Stalingrad. The Anglo Americans land in North Africa.

The tide of war rapidly turns with the enemy forced onto retreat. Tragically, outstanding South African General, Dan Pienaar is killed in an air crash on his way back to South Africa.

1943: The Year of Recovery for the Allies.

The German 6th Army surrenders at Stalingrad with the loss of a quarter million men. At Casablanca the Allied powers proclaim "Unconditional Surrender," and in May the Afrika Korps in Tunisia surrenders bringing the end to the war in Africa. Many South African troops arrive home on leave to a rousing welcome and Brigadier Evered Poole is promoted to Major General; Field Marshall Smuts wins the South African General election.

The U-Boats are defeated in the Atlantic and the Germans are defeated in the largest tank battle in Russia; the Allies invade Sicily in July and the Italian mainland in September. The air battle over Germany intensifies. Led by Wing Commander Guy Gibson VC of 617 Squadron, the "Dambuster Raid" with the specially designed "bouncing bomb", is a huge propaganda success.

Benito Mussolini is deposed in Italy and imprisoned, only to be rescued by Nazi special troops and flown to Germany; the German Battle Cruiser Scharnhorst in sunk off Norway, and the American "Island Hopping" campaign in the Pacific proves successful.

1944: A Decisive Year for the Allies.

By far the biggest event is the long awaited D-Day landings on 6th June to liberate Europe from the Nazi's. All South African radio stations are interrupted to announce the news.

Prime Minister Smuts visits troops in Italy. In the east the siege of Leningrad is lifted and the Russians regain most lost territory. The Allies, including the South African Armoured Division, become bogged down in Italy and the controversial bombing of Monte Cassino occurs.

Stories of German atrocities to Jews in concentration camps filter through to the West. British forces advance in Burma. An attempt is made on Hitler's life in July.

Paris is liberated in August. The V1 rocket is launched on London to start the second blitz.

Field Marshall Montgomery's planned air drop at Arnhem (a bridge too far) is a disaster.

The V2 rocket is launched. Many South African airmen are killed in the relief of the Warsaw Ghetto, Poland

South Africa's 3rd VC is awarded to Lieutenant (later Captain) Gerard Norton of the Union Defence Force for his valiant attack on German machine gun positions in Italy.

Field Marshall Erwin Rommel commits suicide rather than face trial for the attempt on Hitler's life.
The popular band leader Glenn Miller is killed in an air crash.

The Germans are heavily defeated in the Battle of the Bulge in Belgium; many American prisoners
are murdered.

1945: End of the War.
South African 6th Armoured Division makes rapid gains in Italy; the Allies cross the Rhine; the Russians cross eastern German border. Dresden is bombed, killing an estimated one hundred and fifty thousand civilians. Most concentration camps where six million Jews perished, are liberated.
On his return from an operation over enemy territory, South African Major Edwin Swales, of the elite Pathfinder Squadron is shot down in Belgium and awarded the VC posthumously.

USA President Roosevelt dies. Benito Mussolini is executed by Italian partisans.

Russian and Allied troops meet in Germany. Adolf Hitler commits suicide, and Berlin falls to the Russians.

Germany surrenders unconditionally and there are wild scenes of rejoicing in many South African towns and cities.
Two atomic bombs are dropped on Hiroshima and Nagasaki, Japan, in August to bring an end to the war in the Pacific.

World War Two is over.

FOREWORD

An excellently researched document that is rich in factual detail. Even people who have never felt the need to delve into the history of World War Two would find the information of great interest. The writing is didactic and the facts could be used as mnemonics for the reader to conduct his or her own research.

The author has strengthened the historical detail by analysing the various strengths and weaknesses of some of the generals as well as analysing the strengths and weaknesses of the campaigns themselves. The author has also raised alternatives to the various actions in the form of "what ifs". This brings the reader into the present era, answering questions that the reader would indubitably raise upon reading the historical narratives.

The written style is derived directly from the style used by the author when presenting to a large and very erudite audience. Upon reading this document the reader should bear this in mind, as there might be information offered but not expanded upon due to the fact that the background to the information was taken and accepted by the audience, as read. In other words, the audience had some or even a great deal of knowledge about the event or events being written about. As such there is no prolix which results in concise information.

There is no doubt that the author has a most profound knowledge of the history of World War Two. His research has been prodigious; nonetheless he has taken pains to present the information in a manner to ensure that literally every member of the reading public would derive great pleasure in reading this compendium of information.

The work is a very valuable document which could take its place proudly among other works of World War Two.

Dr. Graham L. Coggin

~ONE~
THE 1938 MUNICH CRISIS

In September 1938 Britain and France in a desperate effort to avoid war, capitulated to Hitler at Munich by agreeing to give part of Czechoslovakia to the Third Reich. Hitler, despite this agreement, took the rest of Czechoslovakia in March 1939. The policy of appeasement championed by British Prime Minister Neville Chamberlain lay in ruins.

The Munich Crisis arose from Hitler's territorial claims to regions with large German populations; an issue for which he was prepared to go to war; a war that the Anglo-French were determined to avoid at all costs. They were later to be derided for pursuing a policy of "Peace at any price".

Czechoslovakia was created out of the 1919 Treaty of Versailles. In hindsight it was inevitable that trouble would evolve between the various ethnic groups. This was especially true of the Sudeten Germans who resented living under the rule of foreigners; and demanded that the region be transferred to Germany. Hitler exploited this situation and increasingly made inflammatory speeches; declaring that Germans in Czechoslovakia be united with their homeland.

He demanded that all Germans be incorporated into one Reich, stating that no German national should have to live outside Germany. Consequently, in 1938 he ordered his generals to plan for the invasion of Czechoslovakia. War seemed more and more likely.

The French government tried to forestall Hitler by declaring it would go to the aid of Czechoslovakia in the event of a German invasion. Realising the gravity of the situation, and desperate to solve the crisis without war, Chamberlain proposed to meet with Hitler.

Three meetings were to take place. At the first meeting in Berchtesgaden on September 15th 1938, Hitler agreed not to take any military action against Czechoslovakia. Chamberlain agreed that Sudetenland should be ceded to Germany and promised to persuade the French and Czechs to accept this decision.

The second meeting took place in Godesberg on September 22nd. Chamberlain expected the agreement to be formally signed by Hitler. Instead, Hitler presented Chamberlain with an "Ultimatum". Sudetenland was to be occupied by the German Army immediately and all Czechs evacuated. This was tantamount to a declaration of war and rejected by the British and French. The Czechs relied on military assistance from the French, but, Hitler knew the French would do nothing.

At this time the RAF was undergoing conversion from bi-planes to monoplanes which were not ready for combat. After Munich the Hurricanes and Spitfires that were later to win the Battle of Britain came into production. Chamberlain had a further dread; his military chiefs advised that over one million civilians would be killed by bombing raids in just two months. Mass graves would be needed and there was not enough wood for coffins. Taking all of this into consideration, it is not surprising that many people supported Chamberlain's efforts to achieve a negotiated settlement.

At the second meeting Britain and France felt compelled to prepare for war. On 23rd September, the Czech government ordered a general mobilisation; more than one million fully equipped troops were ready to repel a German attack. On 24th September, the French ordered a partial mobilisation, their first since World War One. Britain mobilised the Royal Navy, Europe was on the brink of war.

Mussolini intervened as mediator in an attempt to avoid war. He proposed that a four-power conference be convened immediately to settle the dispute.

Mussolini persuaded Hitler to hold a third meeting, and on 29th September Chamberlain flew to Munich. Germany, Britain, France and Italy were represented - Czechoslovakia was not. Neither was the Soviet Union, which greatly angered Stalin. Without consulting the Czechs, the four powers agreed that the Sudetenland should be given to Germany immediately.

The Munich Agreement, as it was known, was agreed to by the Western Powers on Hitler's terms. The German army was to complete the occupation of the Sudetenland by 10th October. It was almost identical to the "Ultimatum" submitted by Hitler to Chamberlain at the previous meeting. Such was the desire of the Allies to maintain peace at all costs and their total lack of resolve to halt Hitler's aggression. The Czech Government was informed by Britain and France that it could either resist Germany alone or submit. So much for the French assurance signed six months earlier, to go to the assistance of Czechoslovakia if invaded by Germany. Hitler had been proved right.

After the agreement was signed, Chamberlain went to Hitler and asked him to sign a peace treaty between Britain and Germany. Hitler happily agreed to sign what he later derided as Chamberlain's "Scrap of paper".

Chamberlain returned to London as a hero; triumphantly delivering his famous "Peace in our time" speech to ecstatic crowds. Imminent war had been avoided and Chamberlain was recognised as the man who had saved Europe. However, there were other reactions; the First Lord of the Admiralty, Duff Cooper, resigned in protest from the Cabinet. The Labour Party together with what was regarded as a diehard and reactionary element of the Conservative Party also opposed the agreement.

However, the Munich Agreement was enormously popular in the country for seemingly averting war, and a wave of perverse optimism swept across the country. Chamberlain; announcing on his return that in his view the settlement of the Czechoslovakian problem, which has now been achieved, is only the prelude to a larger settlement in all Europe. Further stating, "This morning I had another talk with the German Chancellor, Herr Hitler, and here is the paper which bears his name upon it as well as mine. I believe it is peace for our time". Chamberlain actually believed that Hitler had been sincere.

Winston Churchill, on denouncing the Agreement in the House of Commons: "We have suffered a total and unmitigated defeat. It is a disaster of the first magnitude; we have sustained a defeat without a war".

Adolf Hitler stated in a speech to his generals; "Our enemies are little worms, I saw through them at Munich. Now Poland is in the position I desired. I am only afraid that some fool will again present me with a mediation plan at the last moment". He was determined to have his war.

On 15th March 1939 Hitler invaded Czechoslovakia and drove through Prague in triumph, thereby, dismissing claims he only wanted to bring Germans into the Reich. Chamberlain felt betrayed by the Nazi seizure of Czechoslovakia, and finally realised his policy of appeasement towards the dictator had failed. He immediately began to mobilise the British armed forces on a war footing. France did the same.

In April 1939 the British gave a public guarantee to Poland that were she to be attacked by Germany, Britain would go to war. Stalin, visibly upset by the results of the Munich conference, formed the opinion that the West had actively colluded with Hitler. This belief led the Soviet Union to alter its foreign policy towards Nazi Germany that eventually led to the signing of the Non-Aggression Pact in August 1939.

Although the Chamberlain Government had allowed Czechoslovakia to be abandoned and dismembered; it is claimed that the Munich agreement had bought Britain nearly a year to rearm. However, how accurate is this? It is a known fact that the subjugation of Czechoslovakia denied the Allies the Czech army consisting of thirty seven divisions already mobilised. The Skoda works was the second most important arsenal in Europe that produced equivalent to the entire British armaments output.

Even more disastrous was the alteration in the relevant strength of the German and French armies. With every month that passed from Munich, the German Army not only increased in numbers but also in the quality of training. The French Army meantime suffered from a sapping of morale. The year's breathing space said to be gained by Munich left Britain and France in a much worse position compared to Germany's. In 1939 Nazi Germany's war expenditure was treble that of Britain and France combined. Finally there is the staggering fact that in a single year Hitler had annexed and brought under his control, seven million Austrians and four million Czechs. The balance had indeed turned in his favour.

Britain and France now faced the might of Nazi Germany whose armed forces were at the highest pitch of readiness for what was about to be unleashed. Due to the lack of resolve and pacifist attitudes of Allied Leaders, and the missed opportunities to call Hitler's bluffing, the world once again drifted towards a world war. Munich was the final opportunity to stop the approaching catastrophe that resulted in the death of some fifty million people.

The belief that security can be obtained by throwing a small state to the wolves is a fatal delusion.

~TWO~
OUTBREAK OF WORLD WAR TWO

World War Two was the most terrifying reality of modern times. It was the first global conflict to be fought with equal intensity in all parts of the world. It was fiercer and more destructive than any in history with domination of the world at stake.

Possibly, most people see the Second World War in rather simplistic terms. Hitler, Mussolini and Tojo started the war; they were the aggressors; and got exactly what they deserved. However, in history there are few genuine villains and even fewer genuine heroes.

Today, one can afford to take a less emotional look at the greatest war in all history and examine its causes with a more critical eye. This is mainly due to documents only recently being made available which cast new light on the once seemingly simple story of the good guys defeating the bad guys. It can no longer be argued that Hitler, and Hitler alone, caused the war in Europe, any more than it can be argued that Japan started the war in the Pacific. It is debatable that both Germany and Japan were provoked by the Western Powers.

The Germany of the Weimar Republic was a creation of the First World War; it was born out of defeat and humiliation. By 1933 the Weimar Republic was wholly discredited, blamed both for the Treaty of Versailles, that shackled Germany, and for the Great Depression.

Few Germans accepted the events of the final months of the First World War.

After four nightmare years, they would not believe that the German army had been beaten. After all Germany's frontier had not been crossed by enemy troops in action. Indeed, only a few months before the wars end, the German army had pushed back the Allied front to within forty miles of Paris. At that time, to the German civilian, victory seemed imminent. Then, inexplicably, General Ludendorf urged the Berlin government to make peace. The German public was shocked when on the eleventh hour on the eleventh day of the eleventh month the armistice to end the First World War came into effect in 1918.

Bismarck's second Reich had collapsed, betrayed by the so called "November criminals". To the masses, the German armies were not defeated in the field. They marched back into their own country with flags flying and bands playing after their long occupation of other countries, only to find that sailors and civilians had attempted a revolution whilst they had been fighting. The victorious powers then proceeded to strip Germany of territory and demanded payment of huge reparations.

This served Hitler and the Nazis purpose and they promoted the "stab in the back" propaganda. The Treaty of Versailles imposed upon Germany undoubtedly sowed the seeds of World War Two. Germany sought vengeance and this was given added stimulus when Hitler came to power in 1933. Hitler threw off the shackles of the Versailles Treaty, remilitarised the Rhineland, introduced conscription, annexed Austria and occupied Czechoslovakia. Hitler promised and provided jobs; he also eradicated most political rights and began harassment of the Jews.

Britain and France did not begin to stir until Hitler's troops re-occupied the Rhineland in March 1936. At this stage France alone had the capability of moving into the Rhineland and if she had done, so, Hitler would have backed down rather than face a confrontation. But France lacked the will. From this moment on, Hitler knew that the Western Democracies could be pushed. Only France believed in upholding the letter of the Treaty. Britain had no interest in the

Treaty, which she now considered unjust (in theory anyway).

When Hitler marched the Wehrmacht into Austria in March 1938, Britain and France merely complained and did nothing. As long as he confined his interests to Eastern Europe, they were more or less content to let him have his way. The annexation of the Sudetenland of Czechoslovakia was another matter. But Hitler argued that the people there were German-speaking and wanted union with Germany, which was true.

British Prime Minister, Neville Chamberlain had no intention of going to war, and the Munich Conference sealed the fate of Czechoslovakia. It is now known that if Britain and France had resisted, the German General Staff would have overthrown Hitler. But the Western Powers had allowed Czechoslovakia to be abandoned and dismembered. On 15th March 1939 in the aftermath of Munich, Hitler invaded Czechoslovakia and drove through Prague in triumph. Thereby, dismissing claims that he only wanted to bring Germans into the Reich. Millions of Slavs were unwillingly wedded to Germany, and Hitler's long-term goal of a German empire in the east was now evident.

This was a decisive moment for the West. Now it could no longer be argued that Hitler only wanted to bring Germans in Central Europe into his Reich. Herr Hitler was no longer considered a gentleman to be trusted. The policy of appeasement championed by Neville Chamberlain had failed. He felt betrayed by the Nazi seizure of Czechoslovakia. The will of Britain stiffened and with it the resolve of France, who was pledged to defend most of the states of Eastern Europe. The British began to mobilise on a war footing. The French did the same.

With Austria and Czechoslovakia now in his hands, Hitler turned to Poland. It was intolerable to him that the German province of East Prussia on the Baltic was separated from the rest of the Reich by a corridor that gave Poland its only access to the sea at Danzig.

When Hitler began his demand for the return of German Danzig, then under the League of Nations authority as a free city, Britain began to talk to the nation that she feared as much as Germany, the

Soviet Union. But Stalin saw at once that Britain's tentative steps towards a Russian alliance were not sincere, and soon began covert negotiations with his bitter ideological enemy.

Hitler read the situation correctly and initiated a brilliant diplomatic counter stroke which completely outwitted the Western Allies. Neither the Axis powers nor the democracies could engage in war without reaching some prior understanding with Russia. Most importantly, he was in a position to offer Stalin something that Britain and France could not, "half of Poland".

Hitler knew that Stalin wanted a defensive buffer of foreign soil between the Soviet Union and Germany and this could be provided with the partition of Poland. Hitler was aware that the Soviet Union clearly controlled the balance of power in Europe, and moved decisively and speedily to conclude the Nazi – Soviet non-aggression pact. The two political antagonists had joined hands. Germany was now free to attack the west without fear from the east thus avoiding a war on two fronts as predicted in "Mein Kampf".

The war clouds now gained momentum; Hitler had hoped that the pact would have shocked Britain and France into repudiating their pledges to Poland. He boasted that all the Western Powers would do is to recall their ambassadors when he attacked Poland. They would never go to war now that the situation was altered.

He was wrong! War broke out on 1st September 1939 with the German attack on Poland. On the evening of 2nd September Chamberlain addressed the House of Commons. MP's expected to hear that war had been declared. Instead Chamberlain said that if the German government would agree to withdraw their troops from Poland, the British government would forget everything that had happened, and diplomacy could start again. Chamberlain sat down in dead silence. The Prime Minister had merely expressed his grief at Poland's fate, not promised direct action. The House was outraged and Chamberlain was warned that he would be out of office if war were not declared. The very men who had applauded Munich now insisted on war. The cabinet met late at night and resolved that an ultimatum should be sent to Germany. The British ultimatum was delivered in Berlin at 9 a.m. on 3rd September. The German government made no reply, and the ultimatum expired at 11am. War was declared. The French followed suit at 5 pm.

What all Europe had dreaded for the previous twenty years was about to happen again. Unlike in 1914, the outbreak was not greeted with patriotic hysteria. People could remember only too well the awful carnage and were fearful of the threat of mass aerial bombing. In 1939 the situation was accepted soberly, with the realisation that Hitler was a cancer within Europe which must be eradicated before peace could be achieved.

The German blitzkrieg in the east coupled with the sitzkrieg in the west, quickly subdued Poland, especially after the Russians in accordance with the non-aggression advanced into Poland. At sea the situation was very different. Here there was no sitzkrieg. On the very first day of hostilities, the liner 'Athenia' was sunk with large loss of life. Two more early disasters were encountered when the aircraft carrier 'Courageous' was sunk. This was followed by the sinking of the battleship 'Royal Oak' at Scapa Flow the following month.

The priority of the army was to transport the BEF troops across the channel. This was duly accomplished under the command of General Lord Gort, a distinguished guardsman who had been awarded the VC in World War One. More than one hundred and fifty thousand men, vehicles, and supplies were moved to bases on the Franco Belgium border.

Was the outbreak of war really necessary? It is possible that Hitler intended to conquer Europe at some time. It is also possible that the British government intended at some time to resist him. But it was neither of these intentions that caused the actual outbreak of war. In September 1939 Hitler merely wanted Danzig and the corridor, and Chamberlain wanted to give them to him. These plans were wrecked first by Polish obstinacy and then by the indignation of the House of Commons.

It is a fact that most countries were pushed into war. The Poles had no choice. The French were dragged along by the British. The Russians and Americans, both mighty boasters, declared it was a crusade against Fascism. However, they had waited until attacked. Only the British and Empire people went to war of their own free will. They were not concerned about Fascism.

Although Britain and France reluctantly declared war against Germany in September 1939 they did not wage it in defence of Poland. They went to war out of national pride and for the sake of national honour and ultimately played a major role in bringing Hitler down. There are many lessons to be learned from the Second World War, but perhaps the most important is this: The horrors of cities destroyed, mass murder and genocide, atomic warfare, concentration camps, death marches, starvation, occupation and human degradation of an unparalleled nature, could all have been avoided.

In order to avoid the tragedy of the First World War, the Western Powers tried to ignore what was happening in Europe and Asia until it was almost too late. The legacy of appeasement which itself was the legacy of the First World War hung like a pall over the West. That war seemed to achieve nothing, and with so many millions of lives lost, it seemed insane to repeat the error. Even as late as May 1940 there were still those in Britain who favoured a policy of accommodation with the new German order. However, the appointment of Churchill rather than Lord Halifax as the replacement for Churchill finally put an end to any thought of appeasement.

It was this disillusionment and horror of war that blinded the West to the threat Hitler posed. By trying to pretend that aggression was not really aggression, that events in far-off countries about which they 'knew nothing' did not concern them. By callously ignoring human suffering, the Western democracies themselves were forced to suffer.

It can only be hoped that this lesson, lost on the generation who did not want to repeat the First World War, is not lost on the generation which does not want to repeat the tragedy and horror of the Second World War.

~THREE~
THE BATTLE OF THE RIVER PLATE

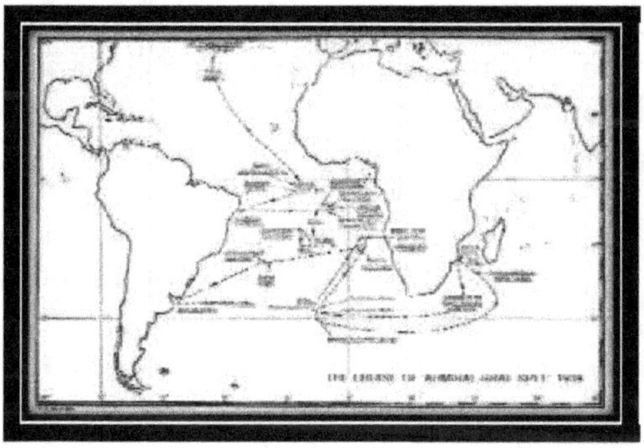

On December 1939 a Royal Navy Squadron consisting of the Heavy Cruiser Exeter, and the Light Cruisers Ajax and Achilles intercepted the German Pocket Battleship Graf Spee. A classic naval engagement then took place off the South American coast near Uruguay that became renowned as the Battle of the River Plate. During the battle the Graf Spee put the Exeter out of action, seriously damaging both the Ajax, and the Achilles.

The Graf Spee, however, also received a number of hits and her captain thought it necessary to take refuge in Montevideo for repairs. He was convinced that stronger British naval forces were close at hand and being unable to complete the repairs within the allotted time, the Graf Spee was blown up by her crew.

This spectacular feat of naval arms, won primarily by psychological means, held the world's attention. It also earned worldwide admiration for the Royal Navy and gave a lift to British morale. The destruction, of such a formidable warship that the Germans claimed to be invincible, by three outgunned British cruisers, set off a great outburst of rejoicing in Britain. It was the first victory of the war and Churchill summed up the mood of the people when he exulted; *"In a cold winter it warmed the very cockles of the nation's heart".*

Why did this naval battle mean so much to the people of the British Empire? Particularly, as it virtually paled into insignificance when compared to later epic sea battles, such as the sinking of the Bismarck; the Fleet Air Arm victory at Taranto that crippled the Italian navy; Pearl Harbour that brought the United States into the war; Midway, that broke the back of the Japanese naval air strike force; and Leyte Gulf that destroyed the Japanese surface fleet to establish the United States as an incontestable naval power. All of these operations were of far greater importance than the Battle of the River Plate.

There are several factors that need to be considered; at the end of World War One, under the Versailles Treaty, Germany was only permitted to build warships of a certain tonnage and armament. Battleships in particular, were restricted to a mere ten thousand tons. However, due to German ingenuity and technological skills, there emerged a completely new warship design and this unique concept was dubbed the Pocket Battleship. These brilliantly engineered warships had six 11" guns plus eight 5.9" guns and were capable of 26 knots. They were slightly larger than the Washington Treaty conventional cruisers but much smaller than contemporary battleships. These diesel powered warships had a wide radius of action that allowed them to cruise over the oceans without relying on land bases.

Three of them were built; besides the Graf Spee, there was the Deutschland and the Admiral Scheer, each with a complement of nearly one thousand two hundred men. The Deutschland was later renamed Lutzow as Hitler did not wish any ship named after the Fatherland to be sunk. These ships were expected to play a crucial role in the "Battle of the Atlantic". A campaign that would totally dominate British naval policy throughout the war; on which everything ultimately depended. The sacrifice was horrific; two thousand eight hundred Allied merchant vessels, and almost two hundred warships were sunk, culminating in the loss of forty thousand Allied seamen and fifteen million tons of shipping.

Focusing specifically on the Graf Spee; the saga opened when she slipped out of port in August 1939, just before the outbreak of hostilities; secretly sailing to the South American shipping routes. Graf Spee's supply ship the "Altmark" also sailed to a predetermined rendezvous position in the Atlantic. The German naval command hoped to achieve immediate and crushing results through the operation of their pocket battleships.

On the declaration of war on 3rd September 1939, Graf Spee camouflaged her appearance and successfully deceived merchantmen into thinking that she was a French heavy cruiser. She also proved to be most elusive, and her speed and unpredictability enabled her to sink nine merchant ships totaling fifty thousand tons in three months. Some of Graf Spee's victims did, however, manage to transmit distress signals, thus alerting the admiralty that a pocket battleship was at large. As a result all available Allied battleships, battle cruisers and cruisers were formed into powerful hunting groups to search the entire Atlantic oceans from Greenland to the Falklands. This was of course part of German strategy; that is to disperse the Royal Navy's superior strength.

With this intention, the Graf Spee headed for the Indian Ocean, and on 3rd November she steamed four hundred miles south of Cape Town. Well out of reach of air reconnaissance; she thrust up the South African coast line to the Mozambique Channel. Whilst on route consideration was given to using Graf Spee's Arado spotter aircraft to bomb Durban's oil storage tanks, but this was considered too risky and abandoned. Shortly after arriving in the Mozambique Channel, the Graf Spee intercepted and sank the British oil tanker "Africa Shell".

Consequently, a senior officer stationed at Durban, on receiving a distress signal from "Africa Shell" radioed a warning of the presence of a German surface raider in the Indian Ocean. This electrified the admiralty, resulting in several Royal Navy hunting ships being redirected to this area. Graf Spee's Captain Langsdorff picked up the warning message and concluded that his mission in the Indian Ocean had succeeded. He therefore returned to the Atlantic for a final campaign before triumphantly returning to Germany for Christmas and an engine overhaul.

Meantime, Commodore Henry Harwood, commander of the Royal Navy cruiser squadron "Force G" in the South Atlantic had shrewdly calculated the reason behind the Graf Spee's foray into the Indian Ocean. He estimated she would soon be lured back to the South

Atlantic and the rich shipping harvest at the River Plate. He, therefore, patrolled this region with the heavy cruiser Exeter and the light cruisers Ajax and Achilles. He was aware that the Graf Spee had the advantage of knowing that all warships sighted were enemy ships and due to her taller look-out mast and aircraft would sight the Royal Navy ships first. This is precisely what occurred shortly after sunrise on 13th December, when the Graf Spee lookout spotted warship masts on the horizon.

Langsdorff knew they had to be enemy ships, but due to his spotter plane being out of action, he mistakenly assumed they were probably light vessels escorting a convoy. He soon realised his mistake when the heavy cruiser Exeter accompanied by two light cruisers were identified. Nevertheless, he reasoned that he would be unable to shake off the faster British ships and decided that action was unavoidable. But in committing the Graf Spee to attack, Langsdorff was effectively ignoring strict orders not to engage enemy warships. He was specifically forbidden to expose his ship to the risk of a naval battle. In total disregard of instructions, Langsdorff ordered battle stations and full speed ahead towards his adversaries.

Commodore Harwood aboard his flag ship Ajax ordered captain Bell on the Exeter to close and investigate the smoke on the horizon. The pocket battleship was quickly identified; at long last the Royal Navy had found its elusive enemy and the "Battle of the River Plate" was about to commence. Ajax, Achilles and Exeter prepared to engage Germany's fabled pocket battleship in deadly combat and more than one hundred brave young sailors would lose their lives.

In terms of weight of guns and armour the odds were certainly in German favour, but the British had the advantage of numbers and manoeuverability. The Graf Spee's gunners were now forced to choose between three objectives while the British could concentrate on only one target. Graf Spee's 11 inch guns had a range of seventeen miles, the 6 inch guns of Ajax and Achilles, ten miles and Exeter's 8 inch guns, fifteen miles. On the face of it Langsdorff had little to fear, and could engage effectively out of range of the three British cruisers. He was confident that his much more powerful main armament would make short work of the Royal Navy's lighter ships.

On the other hand and regardless of the odds, Harwood had a centuries old naval tradition to uphold, and immediately deployed his ships into their pre-arranged positions with the objective of splitting the enemy's fire. He likewise ordered action stations and full speed ahead; the heavy cruiser Exeter to engage from one side and the two light cruisers from the other. The German pocket battleship prepared to simultaneously engage the two smaller ships with her secondary guns and trained her 11inch guns on the Exeter. Visibility was near perfect when Graf Spee closed rapidly and opened fire. In deciding to go for a quick kill, Langsdorff misjudged and this cost him his principal advantage of not only out-ranging but also out-gunning his adversaries. But, it was too late for that now.

The Graf Spee concentrated on the Exeter and soon straddled her to wreck "B' turret. The next salvo hit Exeter amidships and everyone on the bridge except Captain Bell was badly injured or killed. It is difficult to imagine what it must be like aboard a ship during battle. There is little to compare to the hell and brutality of a sea-battle; men are trapped in combating warships that literally throw tons of high explosives at each other, and when hit, a deafening explosion sends lethal particles of shrapnel within the ship's interior; causing unfortunate men to be cut down instantly in death or mutilation. The fire, fumes and flooding in darkened confined spaces must bring terror to the survivors.

Returning to the action, as Graf Spee closed in to finish off the Exeter, her sister ships, Ajax and Achilles raced forward with all guns blazing, forcing the German pocket battleship to switch her heavy armament to them. Thus allowing Captain Bell to take the badly listing Exeter away just as her last gun was put out of action. She had been reduced to a floating inferno and forced to limp southwards to the Falklands.

Most of Ajax's guns were soon out of action, and under cover of a smoke screen fired torpedoes at her foe. Graf Spee turned away to avoid the torpedoes and at this point, inexplicably seemed to lose heart and decided to break off the ninety minute action. It is difficult to imagine why a much heavier armed and formidable ship should run from two damaged light cruisers instead of eliminating them.

The difference was possibly due to the British commander knowing exactly what he intended to do and the German commander not. In deciding to attack, Langsdorff should have finished off the Exeter quickly before the two light cruisers could take up position. By frequently changing from one target to the other, the Graf Spee's rate of fire was enormously slowed up. Possibly, it was Harwood's aggressive tactics that confused Langsdorff. This combined with his indecisiveness over which target to engage with the main armament, the Exeter to starboard or the Achilles and Ajax to port cost the famed pocket battle ship the battle.

Casualties on the British side, however, were heavier than on the German. The crippled Exeter had lost sixty four of her officers and men. The badly damaged Ajax had seven dead and there were many wounded on the Achilles. Graf Spee's damage initially appeared superficial, but inspection revealed that the galleys were wrecked, and some secondary guns had been put out of action. She was also holed near the waterline, and although she had taken twenty hits and lost thirty six of her crew, her fighting ability was unimpaired.

Langsdorff now had immediate and crucial decisions to make; he ultimately came to the conclusion that his ship was not sufficiently seaworthy to reach Germany. Therefore, he decided to make for the shelter of the nearest neutral port of Montevideo to patch up the damage. Captain Langsdorff had been wounded twice during the action and knocked unconscious; perhaps the temporary concussion he had suffered affected his judgment in reaching this decision.

Several further exchanges of fire were made between Graf Spee and the two damaged British ships trailing her until she docked at Montevideo. On arrival, Captain Langsdorff greeted Herr Otto Langmann, the German Minister with a smart naval salute. The Minister replying with the Nazi version ruefully commented, "Gentlemen I wish I could say welcome to Uruguay, but you have made a serious error in bringing your ship here". The reason was, although Uruguay was a neutral country; her sympathies lay with the Allies. Therefore, the international law that states belligerent war ships are only entitled to stay in a neutral harbour for twenty four hours was imposed. The Uruguayans later extended the deadline to seventy two hours but refused any further extension. Strenuous political negotiation failed to gain further time and Graf Spee was ordered to leave neutral waters by 8pm on Sunday.

The German authorities actually requested two weeks to repair Graf Spee's damage in the hope that this would provide sufficient time for U Boats to reach the scene and assist the pocket battleship. Ironically, the British also wanted the departure delayed, to provide time for heavier Royal Navy units to arrive. Meanwhile the crew of the Graf Spee was allowed to disembark in full uniform for the burial of the thirty six fellow crew members that had been killed during the battle. With the exception of Captain Langsdorff, everyone at the service, including the priests, saluted in Nazi style.

During the next few days intense British diplomatic manoeuvres combined with false and misleading reports led Langsdorff to conclude that he was trapped and internment or scuttling were the only choices open to him. He believed that Royal Navy heavy ships including the Battle Cruiser Renown and the Aircraft Carrier Ark Royal were waiting for him. The Graf Spee had used up more than half her ammunition during the battle. He reasoned, therefore, that a similar naval battle would have been beyond Graf Spee's fighting capacity. He deemed the situation hopeless and erroneously thinking that a powerful British fleet was off the coast, communicated his position to the German High Command and put forward three alternatives for evaluation:

> *Internment in Montevideo after the seventy two hour deadline had lapsed.
> *Fight out to open sea, or;
> *Scuttle the ship.

German High Command negated any idea of internment and left the final choice to Captain Langsdorff between fighting out to sea and scuttling the ship with the proviso that if he chose to scuttle, then he must ensure the effective destruction of his ship. Captain Langsdorff addressed his crew informing them that he was not prepared to engage in a senseless battle that would only serve to sacrifice their lives in a death or glory attempt to break out to open sea. He further explained that if he chose to ignore the deadline, his ship would certainly be interned and classified information may fall into British hands.

Consequently, on 17th December the Graf Spee set off slowly, with battle ensigns flying, from Montevideo harbour toward her fate. The Royal Navy ships reinforced by the heavy cruiser Cumberland that had steamed at record speed from the Falklands closed up for action stations. Over a quarter million people had gathered on the waterfront eager to witness a great naval battle.

The international spotlight for the previous four dramatic days had been centered on Montevideo. Front pages around the world were reporting the full story and an American radio broadcast, live from Montevideo, filled the international airwaves with updates of the developments. The British cruiser squadron and millions of anxious listeners expected the pocket battleship to come out with guns blazing. Naval tradition demanded she battle her way out or go down fighting. Tension mounted as Graf Spee made for the territorial limit.

The hundreds of thousands watching the spectacle and the world wide listeners to the running commentaries from Montevideo were shocked when it was sensationally broadcast that a huge explosion had engulfed the Graf Spee. Langsdorff had deduced there was no alternative against reportedly overwhelming odds but to scuttle. This decision brought a fine ship that had sunk fifty thousand tons of merchant shipping, and tied down half the British fleet for three months to an ignominious end. Langsdorff then took his crew on tugs to a German merchant ship and disembarked in Buenos Aires. They were interned and remained there for the rest of the war; many of them stayed on after the war.

It later emerged that Langsdorff had been dissuaded by his officers from personally setting off the explosives and going down with his ship. However, now that he considered he had done all he could possibly do for the welfare of his crew and not wishing any dishonour of the flag, he decided he could not survive his ship. Three days later Captain Langsdorff dressed in full uniform, wrapped himself in the imperial German flag that he had fought under at Jutland, and shot himself. German and British seamen plus local dignitaries paid their last respects to Captain Langsdorff when he was buried with full military honours.

The aftermath of this classic naval engagement and some of the foremost personalities involved, deserve closer examination.

What kind of a person was Captain Langsdorff, bearing in mind that his significant role in the dramatic saga of the Graf Spee has remained controversial and obscure? Hans Wilhelm Langsdorff was the eighteen year-old son of a Düsseldorf judge when he joined the Imperial German Navy in 1912. He saw front line action at the Battle of Jutland in 1916 and served the remainder of the war commanding minesweepers. After the war he served as a Staff Officer before taking command of Graf Spee in 1938. Langsdorff was actually considered by many to be an officer and a gentleman of the old school.

After his death, several noted historians considered him to have been a "first class person". Describing Langsdorff as a highly trained, intelligent naval officer who achieved his wartime objectives while maintaining personal codes of honour and decency; faithfully fulfilling his duties. They further state, *"In the Second World War, mankind sank to abysmal levels of inhumanity. But, in December 1939, German Captain Hans Langsdorff gave the world a matchless example of personal integrity and human compassion"*.

So, it would appear that at the time of the Battle of the River Plate, he was thought to be an exceptional naval commander and a man of the highest character. This may be borne out by the fact that he had dispatched Allied merchant shipping without inflicting the loss of a single life, even though this certainly put his ship and crew at risk. Of the sixty two prisoners from captured merchant vessels on board of the Graf Spee, not one got harmed, not even during the battle.

But in Nazi Germany the media information about the battle was suppressed and Graf Spee's commander never received any credit for his efforts. In fact, the fatherland he fought and died for demeaned his actions. Hitler was most displeased with Langsdorff and criticised him for not fighting to the finish and going down with his ship. Immediately after the loss of Graf Spee, Admiral Raeder, the German Navy Chief, also criticised Captain Langsdorff, claiming that he had lost Graf Spee when he ignored standing orders, that is not to seek battle with enemy warships.

Raeder ordered there would be no repeat of Graf Spee's scuttling stating: "In future a German warship and her crew are to fight with all their strength until they are victorious or go down with their flag flying". It would appear that this may have been a political standpoint because privately Raeder sent a letter to Langsdorff's mother, praising her son as an excellent officer, remarking favourably on his noble character and stated that he fought like a gentleman and died like a gentleman.

These are all the accolades that were heaped on Captain Langsdorff. But in probing a little deeper, the obvious question is, did he commit suicide to show that he had not acted out of cowardice but to save his men, or, was it to avoid court martial for disobeying orders? Why did he not use his Arado aircraft, or, on the assumption they were unserviceable, use land based air reconnaissance to ascertain actual British naval strength off the coast? It is fine to be an officer and a gentleman, and it is OK to be mister nice guy and be popular with your men; but the priority is to get the job done.

Churchill said after Dunkirk "wars are not won by evacuation". In this case then, scuttling your ship and blowing your head off, regardless of the excuses, do not win naval battles. When has a Royal Navy commander ever scuttled his ship to avoid contact with perceived superior enemy forces? The opposite in fact has occurred on many occasions. An example is the destroyer Glowworm attempting to ram the German heavy cruiser Hipper at Norway. There is also the armed merchantman Jervis Bay taking on the pocket battleship Admiral Scheer in an effort to protect an Atlantic convoy. After all, war is war, and you provide your enemy with victory on a plate by scuttling your ship and committing suicide. The objective is to inflict as much damage as possible on your enemy not to self-destruct.

It has also been suggested that he was ordered by Hitler to commit suicide, but there is no evidence to support this. Furthermore, Captain Langsdorff did not display good tactical awareness and appeared to be devoid of any battle plan, being reactive not proactive. Perhaps he had no stomach for confrontation.

As the war progressed, the pocket battleships, despite the enormous range provided by their diesel engines failed to live up to expectations. In 1942 for example, in comparison to the Allied losses inflicted by the U Boats the entire German surface fleet sank the equivalent of what the U Boats achieved in just one month.

What happened to the Royal Navy ships that seen action at the River Plate? The Exeter, after surviving the 11-inch guns of the Graf Spee later joined Allied operations at the Dutch East Indies and was sunk by the Japanese at the Battle of Java Sea in 1942. The Ajax served the remainder of the war in the Mediterranean. The Achilles also survived the war and saw action at Normandy, destroying a German pillbox; the only Allied warship to achieve this.

Commodore Harwood was knighted and promoted to Rear Admiral. He was only the second naval commander since Admiral Horatio Nelson to be knighted in battle. He died in 1950.

Both British and German participants in the Battle of the River Plate have met several times in friendly reunions since the war. Veterans still celebrate the comradeship that evolved from the River Plate Battle. For example, a Canadian town in Ontario carries the name "Ajax" in honour of the British cruiser, and has named many streets as a living memorial to the officers and men who manned the British ships. In 1999, a proposal to add Captain Langsdorff's name to the programme was unanimously supported by the veterans. In Germany and Argentina to this day, annual reunions of Graf Spee's surviving crew members honour the memory of Captain Langsdorff and this loyalty to the Captain has not waned in over sixty years despite continued military criticism of his decisions.

Another irony is that the Battle of the River Plate was the opening British naval victory of the Second World War and it took place almost exactly twenty five years after the opening British naval victory of the First World War. The Battle of the Falkland Islands was fought between the same enemies, in the same coastal region. The German commander, Admiral Graf Spee perished with his sons in this engagement with the Royal Navy.

As a postscript, many British merchant crewmen captured by the Graf Spee and later transferred to her supply ship “Altmark”. Whilst en route to prisoner of war camps in Germany, the "Altmark" was intercepted in Norwegian territorial waters by the Royal Navy destroyer "Cossack". On the direct orders of the First Lord of the Admiralty, Winston Churchill, the Royal Navy boarded her and freed the prisoners.

Prior to the "Altmark" incident in February 1940, Hitler had shown little enthusiasm for the invasion of Norway. However, the incident infuriated him and convinced him that Britain was no longer prepared to respect Norwegian neutrality. He subsequently ordered an invasion to be mounted on a top priority basis. A direct result of this campaign was that Churchill replaced Chamberlain as Prime Minister, thus ensuring that Britain would remain in the war until final victory. So, the Battle of the River Plate cannot be viewed in isolation; it did have a profound effect on later developments.

The faint remains of Graf Spee can still be seen in the shallow water and mud where she was scuttled in 1939.

~FOUR~
COLLAPSE IN THE WEST – 1940

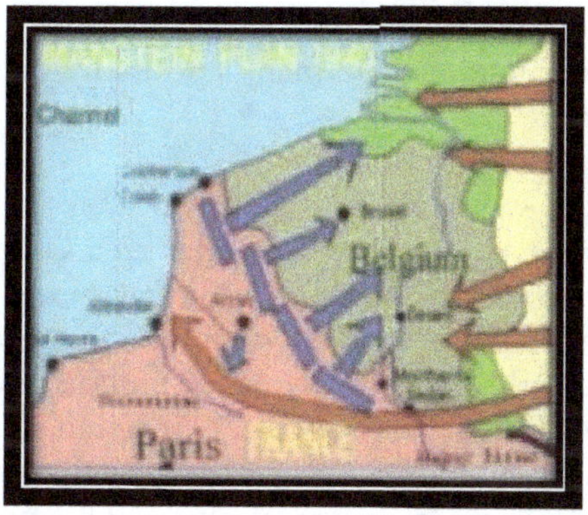

The war to end all wars ended on the eleventh hour of the eleventh day on the eleventh month in 1918. Consequently, there followed, primarily at the insistence of the French, the harsh constraints of the Treaty of Versailles. Defeated Germany was stripped almost bare of all military hardware and the German army was reduced to a token force. In terms of the treaty, Germany was forbidden to build military aircraft, tanks and submarines. The nation that emerged from this terrible war as the strongest military power was undoubtedly France; yet, just over twenty years later the German army astounded the world by sweeping through France with relative ease to inflict a devastating defeat on her arch foe.

Why was this possible, how could this complete reversal of fortunes occur to the nation who had commanded such international acclaim as the foremost military power; why were they overwhelmed so easily? To try and understand the reasons for the catastrophe which befell France in 1940 it is necessary to take a brief look at developments in Europe during the inter-war years.

The Weimar Republic of Germany was a creation of the Great War; it was born of defeat and humiliation, and only reluctantly tolerated by its people, providing the economy was sound. The Wall Street Crash of 1929 followed by the great depression resulted in the demise of the Weimar Republic and the coming to power of Adolf

Hitler in 1933. The new German Führer's message to his people was simple and direct; his avowed aims were;

*To restore prosperity to the population,
*Total German supremacy in Europe, and
*The restoration of Germany's position as a world power.

In the process he would destroy the political system which had, as was popularly believed "stabbed the undefeated German army in the back" in 1918. He served notice of his intentions in his book "Mein Kampf" and he left no one in any doubt that he was determined to settle old scores with France. Hitler almost immediately removed the humiliating shackles of the Versailles Treaty and launched a large scale re armament programme. He expanded the military and Nazi Germany became in effect, a war machine; full employment mainly in the armaments industries led to the country's economic recovery, and the growing popularity of its leader.

France ignored the threat and did not really begin to stir until Hitler's' troops reoccupied the Rhineland in 1936. At this stage France alone had the capability of moving against the small German army; and if she had done so, Hitler by his own admission, realising that he was no match for the French, would have withdrawn rather than risk a confrontation. Although victorious in the First World War; France had suffered devastating losses; she had been bled almost white and determined not to repeat the experience. She lacked the willpower, and her resolve to impose the treaty crumbled. Much complaining and no action were taken against Hitler, who now realised the Western democracies could be pushed.

Hitler alone among the European leaders of the 1930's knew precisely what he wanted to achieve. His opponents, on the other hand, only knew what they wanted to avoid. He, therefore, adopted a more aggressive foreign policy. Britain's Prime Minister Neville Chamberlain was only interested in using the Versailles Treaty to gain concessions in furthering his policy of appeasement. Like the French, he was equally determined that Great Britain should not to be plunged into another war; and believed this could only be achieved by satisfying German grievances. It should be remembered that in those days, as a result of the terrifying air raids which had occurred during the Spanish civil war, everyone believed that aerial bombardment would reduce cities to rubble within a few weeks and European civilization would come to an end. The memory of the Great War still hung like a pall over the Western leaders. The carnage of that war made it appear insane to repeat the same error. The disillusionment and horror of war effectively blinded the West to the threat Hitler posed.

Compared to the new order in Germany, the French nation during the same period had all but lost confidence in itself. In seeing the rise of Hitler's Germany, many Frenchmen viewed the price of victory with bitterness and cynical despair. The mass army had been retained but its effectiveness had stagnated, its usefulness was distrusted and seen as an instrument only capable of bringing about another bloodbath. The doctrine of offensive action was abandoned in favour of a heavily prepared fortified frontier, the celebrated Maginot Line. Successive French governments voted vast sums of revenue for these fortifications covering the frontier with Germany; but, fatally, as it turned out not the frontier with Luxembourg and Belgium in the Ardennes region. The French, with disastrous consequences, were not so willing to spend on their air force which had become seriously neglected. Whereas the military of France had for years suffered from neglect and indifference, the German military had undergone a completely different experience; where professional pride burned as strong as their desire for revenge.

The type of war which Germany was preparing for was very different from that which she had lost in 1918. This time round it would be a lightening campaign named "Blitzkrieg" in which an overwhelming initial blow would settle the issue before the victim would have time to fully mobilise or call on the assistance of allies. When Hitler came to power in 1933, the German military was virtually non-existent, yet, Germany mobilised in 1939 with ten armoured, four motorised, and eighty four infantry divisions. The most powerful battleships in the world, Tirpitz and Bismarck were nearing completion and a new submarine force emerging. The Luftwaffe consisted of almost five thousand; a truly remarkable feat by any standards. Germany invaded Poland in September 1939, throwing France into a crisis; what all European leaders, with the exception of Hitler, had dreaded, was about to happen. However, the Anglo French did honour their commitment to Poland and declared war on Germany. Then proceeded to do absolutely nothing, no plans existed to give Poland direct military assistance.

The Poles had been expected to hold out for at least three months, but, the Germans successfully concluded the campaign in the east within weeks. A hopelessly out-dated and obsolete Polish army quickly disintegrated before a modern armoured force in combination with air supremacy. The French declined to mount any serious attack in the west against the twenty five German divisions. The French army was not geared to the offensive and the small British Expeditionary Force was only to arrive in France four weeks after the declaration of war.

After the conquest of Poland, Hitler offered vague and tentative peace proposals which were rejected; therefore, he ordered that an offensive in the west be prepared. The extremely cautious plan; code named "plan yellow" was devised by the German high command with objectives which were even more limited than the "Schliefen Plan" of the First World War. The German Chief of Staff, General Manstein was unimpressed with "plan yellow' which he considered inappropriate for the concept of Blitzkrieg. Alternatively, he proposed that the main thrust should be made through the Ardennes at Sedan, and then a dash for the channel coast to cut off and isolate the Allied forces, whom he was by this time convinced would advance north into Belgium.

Fortunately for the Germans Manstein's proposal coincided with the news that a German aircraft carrying details of "plan yellow had crash landed in Belgium. It contained sufficient information to persuade the French that as in 1914 the main German blow would come through Belgium. The descent of this aircraft was to prove much more beneficial to the Germans, than to the Allies. Hitler once more postponed the offensive which gave Manstein time to show his Commander in Chief the advantages of his proposal. Hitler, being more imaginative than his senior generals, enthusiastically endorsed Manstein's audacious plan. He also saw the benefit of using paratroops and gliders to capture strategic positions. The main thrust of the German offensive was now to be shifted from the north to the south. The code name for the new plan was operation "Sichelschnitt" - the sweep of the scythe. So, rather than, as in 1914, a right hook, in 1940, it would be a left hook. The prospect now emerged of knocking out France with a decisive blow delivered through the soft centre of her elongated defence lines.

Because of the German invasion of Denmark and Norway, the attack in the west was again delayed until May 1940. The time was utilised in fine tuning the operational plan and in training the troops for the demanding requirements of the Blitzkrieg offensive. German troop morale was high and there was a great deal of enthusiasm for the coming campaign. They were ardent, fuelled by patriotism and had complete faith in their Führer to whom they had all sworn an oath of personal allegiance. Conversely, the picture on the Allied side was very different; French and British troops had been subjected to a long harsh winter, waiting for an enemy who never came. Frequent false alarms caused them to maintain their positions, exposed to the elements. Morale was extremely low, particularly with the French troops in the Sedan sector where the main weight of the German attack was due to descend.

Fifth columnist subversive activities were skilfully used to demoralise French troops, who were subjected to defeatist and anti-British propaganda circulated by French communist party members. Incidents of sabotage in the French armaments industry were frequent and many arrests were made of communists who were loyal to the Soviet-German pact. In 1940 the French army was accepted, particularly by the British, as the most formidable in the world; in reality it was nothing of the kind. It was tactically out of date, and had never really recovered from the shock of the Great War; being politically undermined and tailor-made for disaster.

On paper, the French army was more powerful than the German and had more tanks; amazingly, "some were even better", but they had no conception of massing them for a breakthrough in the manner which the Germans had successfully applied in Poland. The French tank strength was instead spread out thinly and ineffectively; tank organisation was woefully deficient with no clearly defined purpose, and regarded only as useful support for the infantry. The Allies only woke up to the true potential of the tank after the

German victory in Poland and by then it was too late. It was, therefore, essentially a clash of principles between two rival schools. The French believed in distributing their tanks in fairly even proportions along the entire front; whereas the panzers were massed in corps formation, assuring them of local superiority at the point of impact. The German commanders were "tank minded" and fully aware of the capability of the tank; they had calculated the best tactics to gain maximum advantage of this weapons offensive role. These tactics had been practiced on numerous occasions during peace time and in Spain and during the Polish campaign.

The irony of the situation is that it was the British who had first advocated the new theories on mobile tank warfare. Captain Liddell Hart's and General Fuller's writings were avidly studied and eagerly devoured by progressive German commanders and largely ignored by the Allies; with disastrous consequences. Hitler's personal role in the development of this formidable combat weapon was considerable. He had given full support and encouragement to his brilliant commanders, which included Rommel, Guderian, and Manstein.

There had been virtually no military cooperation between Britain and France since the Great War. Indeed the two countries had more often been at loggerheads than in union; resulting in the Allied command structure becoming complex in organisation. They had failed to develop adequate air cover for their ground forces which left their armies horribly exposed to the Luftwaffe. As if this were not enough, the Germans had also broken the French communications code; and therefore, were aware of troop deployments and any supply problems. These shortcomings and incapacities in modern warfare go a long way to explain the weaknesses of the French army's preparations for the campaign of 1940; undoubtedly, the result of pacifist lethargy which had overwhelmed the nation during the inter war years. The British contribution in France led by Lord Gort amounted to a mere ten divisions, (less than half of the Belgian army). A modest air force support was also sent to France; but the valuable Spitfire and some Hurricane fighter squadrons were retained for the defence of Britain. It was in the air that the Allies were at the greatest disadvantage. The remaining Allied forces consisted of, nine Dutch, twenty two

Belgian, and ninety four French divisions, a total of one hundred and twenty five against the German one hundred and thirty six divisions. This then was the situation prior to Hitler striking west.

The war began in earnest and the "phoney war" ended when Hitler struck at the Low Countries and France. On the very same day that the Germans invaded, the British were in the process of changing their prime minister and the French were re-shuffling their government; this was indeed a house divided. The German airborne assault was launched at dawn on May 10th; it was aimed at key sectors in the Low Countries which included the Albert Canal bridges and the fort at Eben-Emael. The Dutch resistance, though very courageous, stood no chance of delaying the Germans seriously, and by May 13th, the situation had become so grave that Queen Wilhelmina and her government were compelled to leave the country. On the afternoon of May 14th, during German-Dutch negotiations for surrender, the notorious air raid on Rotterdam occurred; compelling the Dutch to capitulate.

The Belgian army having fallen back to the defensive positions as agreed with the French anchored their right flank at the heavily fortified complex at Eben-Emael. The Belgian disposition seemed to be reassuring, but no account had been taken of the imaginative flair of Adolf Hitler, who had taken a personal interest in using airborne troops to capture this vital position. The key factor in this daring enterprise was the German troop carrying gliders; they descended silently on their targets landing right in the middle of the Belgian defences to create maximum confusion. The Germans speedily cut the cables to the bridge demolition charges as well as the telephone lines. The highly trained assault pioneers, who had been practicing for months on this operation, systematically neutralised the heavily fortified position which had previously been considered to be impregnable. Rendered helpless by the Germans, the garrison surrendered. The Belgians in other areas continued to fight as best they could, but were unable to prevent the Germans bringing in reinforcements. The Allies launched repeated air strikes at the captured bridges, but the German anti-aircraft and fighter defences fought them off. Most of the Allied planes were shot down and their bombs caused virtually no damage.

The Allied Supreme Commander, Marshall Gamelin, somewhat surprisingly faced the invasion with confidence and set in motion the complicated manoeuvre which was to advance his five armies into the Low Countries after the Germans invaded. However, there were unbelievable snags; his headquarters had no radio transmitter and his army commanders were not all in agreement or fully conversant with his battle plans. The Anglo-French armies proceeded to rush north from defensive positions that had taken nine months to prepare; in order to occupy unprepared positions on the Dyle Line. They then waited to face what they thought was the main German assault. Needless to say, this brought great delight to Hitler and the German high command.

Further south, in the Ardennes region, the German comprehensive assault preparations were being conducted in the utmost secrecy. Entirely fooling Allied intelligence and completely unnoticed; the Germans deployed a formidable force in the Ardennes, ready to strike the exposed, flimsy and strung out French defence lines. The Allied supreme command had taken a great risk in manning this porous and thinned out sector by defending it with untried and unprepared troops. Thus, because they had failed to foresee Hitler's ingenuity, the Allied chiefs were to be caught badly napping. The French military held that this region's narrow roads could not accommodate a large armoured force; even when air reconnaissance reports were received that an armoured build up was in progress, these were largely ignored and considered to be irrelevant. All attention was concentrated to the north; the French command was completely blind to any major threat through the Ardennes. The exceptional logistical achievement of the German general staff, in concealing the attack displayed a remarkable skill in organisation and meticulous detail; they possessed a great talent in tactical planning.

While the Allied armies hurried north to cover what they believed to be the main threat; the Germans launched their daring and what proved to be decisive attack through the Ardennes against the thin crust of defenders. The Ardennes was undoubtedly a formidable obstacle but like any natural barrier it could be conquered if preparations were sufficiently thorough. Rommel and Guderian's divisions spearheaded the advance and broke through the French defences to speedily establish bridgeheads on the River Meuse.

The French defenders in this sector, being reservists and poorly equipped, collapsed almost immediately and were wiped out; thus opening up a huge breach. By nightfall the assault had smashed through into the French rear, causing fearful panic. A general rout developed, with the French fleeing in all directions. The Luftwaffe provided invaluable air support by causing maximum confusion and pulverizing enemy airfields at the precise moment of the ground attack. As the infantry moved forward; Stukas, operating as precision artillery, pounded defensive positions. German aircraft bombed and strafed ahead of advancing columns, smashing roads, blasting tanks, blowing up fuel stores and machine-gunning enemy planes before they were airborne. The Luftwaffe rapidly gained air superiority and inflicted tremendous damage. The Germans energetically exploited their success, and on the following day launched their attack south of Sedan with equally devastating effect. The defenders again broke in confusion, stunned at the speed of the Wehrmacht's advance.

The French army began to lose its coherence, suffering grievous losses in the process. General Gamelin and the French high command believed that the River Meuse could not be crossed without a lengthy build-up of weapons, supplies, and manpower; but the Germans planned this phase of the operation with infinite care. They chose to make their greatest effort over an eighty kilometre front near Sedan against defenders comprising almost entirely of reservists over forty years of age. Everything the French tried to do on May 14th to neutralise the threat at Sedan was too slow and too late. Near suicidal Allied air missions against enemy held bridges were beaten off with frightful losses; the German penetration deepened, scattering the French forces before them. They then consolidated for the race towards Abbeville and the Channel coast.

An energetic Allied counter-attack from the north and south would have had a very good chance of success and may have restored the French front; but, nothing was done. In some sectors the shattering psychological blow simply overwhelmed the front line defenders to create panic, confusion, and disorganised retreat. Few areas managed to coordinate an orderly withdrawal; French resistance was wilting and near to collapse. The Germans continued to probe, seeking softer targets; only occasionally meeting pockets of resistance. Surprise had been achieved on a grand scale and the poorly supported and overextended French lines were taken completely off-guard and off-balance. Constant Luftwaffe air strikes savaged the French supply lines, paralysing traffic, jamming roads, to wreak havoc among the troop columns as well as key railway junctions. An endless stream of panicking refugees flooded the roads to create congestion and bottle necks which further exacerbated military operations.

After their victorious ride through Poland the previous autumn; the Germans had feared that France would be a tougher nut to crack. To their surprise, they found the going easier; the roads were better, and the RAF which might conceivably have stopped them stayed mostly in England. French garages were well-stocked with fuel and Michelin road maps. Poland had never been like this! The French First Army was routed and on the brink of catastrophe; the impact and demoralisation of defeat went right to the top with terrible speed. French Premier, Paul Reynaud, phoned Winston Churchill on May 15th, a mere five days since the German offensive began, to announce, "We are beaten, we have lost the battle; the road to Paris is open."

Churchill flew to Paris the next day to attend a conference; it was his first visit since becoming prime minister. He asked Gamelin in French "where is the strategic reserve"? He was astounded by the reply "there is none". The French high command had gone to pieces and had plunged into despair; the plan devised by Gamelin had gone hideously awry. The Germans proceeded to create a "panzer corridor" to enlarge the breakthrough and consolidate their flanks. But Hitler was now becoming very nervous, and, frightened by his own success; he was ever mindful of the vulnerability of overextended forces and the German reversal on the Marne in 1914. Being far removed from the front, the German high command could not envisage the total collapse of the French army and expressed concern that the leading panzers would be cut off by a counter-attack. The order was given for the German rapid advance to be halted and await the infantry. Guderian was furious and protested vehemently, to the point of offering his resignation. A face saving compromise was reached and he was allowed to continue with "a reconnaissance in force"; the Germans reached the Channel coast on May 20th.

Colonel Charles de Gaulle did lead an armoured counter attack, which had initial success, causing anxiety for the German command; but he had to pull back after being set upon by Stukas, against which he had no defence. The British also launched an uncoordinated armoured attack at Arras to surprise the Germans, but were beaten back by the devastating fire power of the 88mm anti-aircraft guns, which the Germans discovered could be converted to a dual purpose role. For the dazed Allied force, annihilation now seemed inevitable. With disaster threatening, the French Premier Paul Reynaud once more reshuffled his cabinet and recalled the legendary Marshal Petain, the hero of Verdun in World War One; he also sacked Gamelin as Supreme Commander, and replaced him with seventy three year old Weygand, who was summoned from commanding the French forces in Syria. For three days at the height of the battle, there was in effect no Allied Supreme Commander. Meanwhile, one million men were being cut off from the main body of the Allied forces. Even when the Allied forces were cut in two and the elite of their armies effectively isolated, the bewildered French command still believed that the primary objective of the Germans was Paris.

The British Commander in Chief of the BEF, General Lord Gort, on seeing the French First Army disintegrating, and fearing envelopment, made up his mind that there was no point in staying in France. In defiance of French supreme command orders, he decided to retreat towards Dunkirk and establish a secure perimeter.

Hitler once more got the jitters, the British attack on the German extended flank at Arras was precisely what he had feared might happen; this influenced him to issue the notorious "halt order" before Dunkirk. An order that has been the subject of debate almost since it was issued and interpreted in many ways. Some claim that Hitler wished to spare the British the humiliation of total surrender; in the hope that this would make them more amenable to a peace settlement. Others argue that he wished to give Göring, his appointed successor and commander of the Luftwaffe a share of the glory, by allowing him to finish off the trapped British troops. However, perhaps the most credible explanation is that Hitler did not want to commit his panzer forces in the swampy terrain around Dunkirk. He needed to conserve, re-group, and re-fit them for the next phase of the campaign in France. Whatever the reason, it meant that the chance of reaching Dunkirk before the British was missed and the Germans were unable to complete their overall objective; which was the total destruction of all the Allied armies. Meanwhile, the Belgian army finally surrendered on May 28th, which left the British flank wide open and effectively sealed the fate of the BEF in France. With their head now in a noose, they had no choice other than to evacuate to England; by June 4th, three hundred and thirty eight thousand British and French troops escaped in "the miracle of Dunkirk".

There was much rejoicing in Britain about the rescue operation and the media hyped this into a major victory; thus prompting Churchill to announce that "wars are not won by evacuations". This disastrous course of events forced Weygand to abandon his plan for a joint counter-attack against the panzer corridor. He tried extremely hard and with great energy for a man of his age to boost French morale. A defence line was established along the Somme River and orders issued that there was to be no further retreat from present positions. De Gaulle's armoured forces attacked the German bridgeheads along the Somme, but achieved only limited success before the attack ground to a halt; the odds against him were hopeless and his flanks exposed. The unequal struggle continued; on June 5th the Germans launched attacks along the entire Somme front, and for the first time in this campaign the Germans experienced heavy losses, even though they now outnumbered the French by two to one. The French ultimately gave way and the Germans arrived at the gates of Paris on June 12th to trigger off another civilian mass exodus. The Germans, however, had no thought of fighting a costly battle for the capital.

Another development occurred further south to add to France's agony. On June 10th, Italy decided to intervene in the war on the side of her Axis partner; no doubt hoping to snatch a share of the spoils. The Italian campaign was a complete fiasco.

On June 14th the triumphant German army entered undefended Paris which had been declared an open city and evacuated the previous day. The tricolour was lowered and the Nazi swastika flew from the Eiffel Tower. Veterans of World War One openly wept as the Germans held their victory parade down the Champs Elysee. Hitler had achieved the prize which had eluded imperial Germany.

The French commanders were by now discussing how to bring about an end to hostilities in the most favourable terms. Both Britain and France had pledged not to conclude a separate peace with Germany and Churchill refused to release France from her undertaking; Britain he said was determined to continue the struggle no matter what. This embittered the French, considering that Churchill had not released the RAF fighter squadrons to aid them despite their desperate pleas. Opinions are as divided today as they were in 1940; but, there can be no doubt that the refusal to commit the RAF fully to the Battle of France poisoned the relationship between the political and military leaders of Britain and France.

Reynaud wanted to keep the alliance with Britain intact and continue the fight from Algeria; France, he claimed would be able to utilise the resources of the French empire and be supported by the French fleet. In retrospect, this proposal was impractical as the French had virtually drained her colonial army to fight in France. Germany with the assistance of Italy and perhaps even Spain could easily have overrun Algeria where France had no air cover. Faced with these difficulties, the French cabinet gradually came round to concluding that an armistice with Germany would be the wiser option. The appalling fate of Poland which was administered by a Reich's commissioner persuaded many that an armistice would spare the French population from a similar fate.

Being determined to keep France in the war, Churchill belatedly sent Anglo-Canadian reinforcements, and even went as far as to propose a Franco-British union, where both empires would merge and have common citizenship. Reynaud received this proposal enthusiastically but could not persuade his cabinet to accept. He, therefore, resigned on June 16th and was succeeded by Petain who resolved to seek an armistice with Germany.

The major concern for the British was a German take-over of the French fleet, which combined with the existing axis fleet, would prove too strong for the Royal Navy. The French Admiral Darlan promised that this would never be allowed to happen; however, the British insisted that she was only prepared to release France from her obligations on condition that the French fleet sailed immediately to British ports. The French refused; they were by this stage riddled with defeatism and wanted their fleet to be used as a bargaining chip in future negotiations with the Germans. De Gaulle, now promoted to the rank of general, but omitted from Petain's cabinet, decided to leave for London and set up a government in exile; a few weeks later he was sentenced to death as a traitor by a Vichy French military court.

Prior to the cessation of hostilities, Hitler insisted that the French must undergo the humiliation of signing the armistice in the very same railway carriage at Compiegne that Marshall Foch had dictated surrender terms to the Germans in 1918. Negotiations on details of the terms continued for two days, primarily concerning the French fleet. On June 22nd the Germans delivered an ultimatum to the French to sign the documents immediately or hostilities would resume within an hour; the armistice was duly signed. The battle of France was over; the French had lost ninety two thousand men killed and two hundred and fifty thousand wounded. By comparison the Germans lost thirty thousand killed and one hundred and twenty thousand wounded; most of the German losses occurred after Weygand had taken over command. This poses the question, what would have happened if the battle tactics instituted by Weygand had been in force when the campaign began?

The brilliant German operational plan had proved a spectacular success. The German army of 1940 with fresh ideas plus a well organised air support had routed the 1918 French army with 1918 theories. The Germans were masters now of the whole mainland of Western Europe and had reason to strut with pride in an amazing feat of conquest. There were three main reasons for the defeat of the Allies in 1940. The first was that the politicians, both in France and Great Britain, had neglected to prepare their armed services for a modern war. The second was that the majority of the generals were "living in the past", in 1918 to be precise. The third reason was that a vital few of the German generals who had Hitler's ear, had grasped the concept of Blitzkrieg.

France was out of the war and was now at peace, but at what cost? Petain ruled a truncated nation from the small town of Vichy; the Germans occupied all the northern and western parts of the country, including Paris, plus the Atlantic and Channel coast lines. Vichy rapidly became a dictatorship under Petain whose determination to "save" France, developed into full scale collaboration with the Germans. This resolved Churchill to order the Royal Navy to attack the French fleet based in North Africa.

Britain remained unbroken, and now made the largest anti-tank obstacle (the Channel) impassable. Although determined to continue the fight alone; it was extremely doubtful that Britain could hold out indefinitely against the seemingly invincible German war machine. She was, however, as defiant as ever and Hitler's peace offers were rejected; thus, frustrating his plans to end the war. However, he believed that he had won the war, Britain; he claimed, was finished and would soon come to terms, then he would be in a position to deal with Russia.

Why did the British continue the war against seemingly insurmountable odds, especially when Hitler was still prepared to offer generous peace terms? Most sources tend to gloss over this issue and it is only by reading newspapers of this period that the full abhorrence of Hitler's regime is brought home. Some of the clippings read as .follows; "the abominable depths of a gangster leader of a gangster people, who are unscrupulous, treacherous and brutal." "A cowardly nation crazed and intoxicated with the lust for blood".

The German invasion of Britain, "0peration Sea Lion" was planned; but Hitler held no illusions about the logistical difficulties of this formidable task. He knew Germany lacked the means to conquer Britain and the projected invasion was never really a practical possibility. In any case the combined operations required to execute this would inevitably exhaust Germany and allow the Soviet Union additional time to build her military strength. The Germans believed, quite conceivably, that Britain could do nothing effective against Germany while she stood isolated and alone. They felt that she would eventually sue for peace when she realised the hopelessness of her position; especially when she was blockaded by air and sea.
By September 1940 Adolf Hitler had reached the pinnacle of his career. The strategic concept behind the 1940 victory enhanced his military prestige and he was rarely opposed by his generals who hailed him as "the greatest military commander of all time". Only a

few months after provoking the war, he was master of the continent of Europe and was seen to have achieved what no man; including Napoleon had ever done before. He had almost accomplished the impossible at practically no cost. The Wehrmacht was the most highly trained and battle seasoned fighting force in the world with superior skills at all levels. The total German losses in the war to date were insignificant, compared to the casualties in the First World War.

Yet despite all this, Hitler could not bring the war to a successful conclusion! He had played for time with the non-aggression pact with Russia which had been purely expedient, to dispose of the Western Powers and he was not prepared to wait for the balance of power to shift against him. With Allied opposition crushed on the continent he prepared to risk a war on two fronts, contrary to his pledge to the German people.

His thoughts now focused on Russia where he proceeded to snatch defeat from the jaws of victory.

~FIVE~
OPERATION SEALION

Operation Sealion was the invasion planned to take Hitler triumphantly into London. However, it was the invasion that never was. Why was this, why did Hitler not invade Britain; especially after Dunkirk?

To understand Hitler's dilemma it is necessary to go back to the collapse of the French. Even by his own optimistic standards, Hitler had been taken by surprise at the swift fall of France that left Great Britain as his only major Western European enemy.

When France fell in June 1940 Britain was at her most vulnerable and a successful invasion at that point would have ended the war on German terms. Yet, Germany could not capitalise on its amazing good fortune. No contingency plans had been prepared for such an eventuality; and even if they had existed, the Kriegsmarine was totally inadequate for Sealion.

The reasons for Germany's lack of naval readiness to engage the British fleet were entirely political. Hitler never envisioned a long-term war with Britain, much less an invasion. He considered German mastery of the continent to be central to world domination and expected the "nation of shopkeepers" to be sensible and come to terms. Therefore, since Hitler's strategic aims lay on the continent the Wehrmacht and Luftwaffe received top priority. The navy was merely an ancillary service.

Perhaps there was some merit in this as the Blitzkrieg into Poland then the total collapse of France in just six weeks stunned the world. In June 1940 all that stood between the seemingly invincible Wehrmacht and Nazi domination was the English Channel plus some badly shaken British troops recently evacuated from the continent. The shattered remnants of the British army regrouped and frantically prepared, as best they could, to repel an amphibious assault. Britain once more found herself facing a powerful Continental enemy just as she had experienced back in the days of Philip of Spain and of course Napoleon.

On those occasions Britain had resisted invasion by retaining mastery of the Channel. Sea power proved to be the first line of defence, but in 1940 that now needed to be reinforced from the air. The Royal Navy had learned from Norway and France that Sea power was useless if control of the skies wasn't possible. At Dunkirk five of the nine British destroyers lost were due to air attack.

Theoretically, Operation Sealion appeared simple and Britain should have been an easy target. After all no anti-invasion plans had been prepared. British pre-war defence relied, as did the French on the Maginot Line. Events, however, overtook the Allies. The Wehrmacht had achieved astounding success since the attack on Poland and the Luftwaffe had proved to be a formidable force. However, due to the heavy naval losses suffered in the Norwegian campaign, Hitler's operational fleet at the time of Sealion had been reduced to one pocket battleship, four cruisers and a dozen destroyers.

Germany had lost about half its surface navy in the Norwegian campaign, and would have been incapable of keeping the Home Fleet out of the Channel. In effect the British had a 10:1 superiority over the Germans. However, the Royal Navy could not bring this enormous advantage to bear as most of the fleet was engaged in the Atlantic and the Mediterranean, but, as noted, the superiority of the Royal Navy was now challenged by the Luftwaffe. British sea power was no longer the primary issue. It was airpower, and to achieve air superiority the Luftwaffe would have to neutralise the RAF. Only then could British sea power be contained long enough for German ground forces to be ferried across the Channel.

This was the general situation in the summer of 1940. But, how serious was the threat of invasion? Politically, the Führer would have preferred to come to an understanding with Great Britain. He admired the British for the way they had built their Empire and wished to negotiate a peace deal. He was quite convinced that after the defeats in Norway and particularly after the disaster at Dunkirk, Britain would sue for peace. He offered what many considered at the time to be very generous peace terms. Hitler reasoned that a British defeat in 1940 would bring about the disintegration of her Empire and German blood would be shed accomplishing what would only benefit Japan and the United States.

Hitler broadcast his peace offer on July 19th, but Churchill treated this contemptuously, stating Britain would never sue for peace and proceeded to rally public support for his defiant stance. An irate Hitler retaliated; Britain, despite her hopeless situation, and still showing no willingness to come to terms, would be invaded and subjugated. Hitler approved Operation Sealion, announcing that Britain was to be eliminated as a base for future operations against Germany. No detailed plans had been prepared, only a draft.

The British army stationed in Britain at this time consisted of twenty six divisions, twelve of which had only recently been formed and were not fully trained or equipped. The remaining fourteen divisions had lost a vast amount of military hardware in France, thus necessitating obsolete armaments to be stripped from military museums. Under the cash and carry programme Britain was able to purchase half a million rifles and nine hundred artillery pieces from the US. By September 1940, twenty nine divisions, including two Canadian were equipped and available for national defence.

The Royal Air Force had mobilised the last of its reserves and Spitfire production had accelerated. Lord Beaverbrook, the minister responsible for aircraft production appealed to the public to donate scrap metal to build fighters, resulting in mountains of iron and aluminium that mostly could never be transformed into Spitfires or Hurricanes. However, in terms of morale boosting, it was a runaway success.

By this time the German Army staff had submitted an invasion plan that would transport forty one divisions across the Channel. Grand Admiral Raeder, who was not a supporter of Operation Sealion rejected this as impracticable and told Hitler of the difficulty in obtaining landing craft suitable to carry invasion troops. He recommended a limited crossing restricted to the Dover area as the German navy could never provide adequate protection over a wide front. Raeder had a far more realistic view of the difficulties involved in Sealion. He considered that the war would be conducted far more successfully by focusing on the Mediterranean and repeatedly warned Hitler of the dangers associated with a landing in Britain. Particularly when faced by an enemy committed to fight. On the assumption that a beachhead could be established on British soil, Raeder stated there was a danger that the Royal Navy would cut off supplies to German troops, isolate them and force their capitulation.

In 1940 a German infantry division required one hundred tons of supplies per day and a Panzer division consumed three hundred tons. To move nine divisions and sustain them for the first ten days before the second wave was scheduled to land would strain German resources to the limit. In addition to an under strength Navy and inadequate planning, German inter-service rivalries also emerged; especially when the Army saw Operation Sealion as nothing more than a large river crossing in which the "Luftwaffe would do the work of artillery“. The army preferred a broad front to split enemy forces, but the navy wanted a narrow front to facilitate protection of the invasion force. Amphibious combined operations require close cooperation between the various branches and the Germans simply did not have this.

On the last day of July Hitler held a meeting at the Berghof. Raeder detailed why he believed the army plan was untenable and argued for a postponement of the invasion until 1941. All three branches of the German military reiterated the problems associated with an invasion.

It would require;

*Control of the Channel
*Control of the skies,
*Good weather and,
*Destruction of Coastal defences.

The result was a compromise. The invasion force was reduced to twenty seven divisions to provide von Rundstedt with a sufficiently wide front to break out and encircle London. Other groups would head towards Gloucester and Bristol and a feint landing on the Norfolk coast was planned to draw off British reserves. Addressing his service chiefs, Hitler made it clear that he recognised the plan had its dangers; especially those identified by Raeder. But he was keen to press Britain into submission so that he could turn his full attention on his real enemy; Russia.

Hitler therefore, wanted operation Sealion to be over by mid-September. Raeder however claimed the invasion could only start in mid-September providing Göring's Luftwaffe defeated the RAF. As no German battle fleet existed to give off- shore bombardment, long range coastal batteries with ranges of between forty and fifty kilometers would have to be positioned around the Calais area. Combined with massive Stuka attacks, they were planned to neutralise British coastal defences and prevent the Royal Navy from attacking German troop transports.

The army would immediately capture a port in order to land the Panzers. Air supremacy and the early introduction of armour were thus critical to achieve victory. Hitler rejected requests to cancel; if granted, this would undermine the invasion as a political threat. The build-up for invasion had to continue and Britain had to be kept under military pressure.

It was decided that the Luftwaffe should tighten the screw by clearing the channel of British warships and the skies over southeast England of British aircraft. To establish command of the Straits of Dover from mid July the Luftwaffe stepped up the military pressure by attacking the Channel ports and shipping. By the end of July the Royal Navy had to pull all its larger warships out of the channel because of the threat from German aircraft. All seemed to be going to plan; this mounting military pressure and the prospect of invasion were intended to break British spirits and render Operation Sealion unnecessary.

Certainly, neither the threat of imminent invasion nor offers of an 'honourable' peace had done the trick. It appeared that Germany would actually have to execute one of the most difficult military operations imaginable; an invasion, launched across at least thirty five kilometres of sea culminating in a landing on a fortified and desperately defended coast line. It was immediately clear that this could not even be attempted until the Royal Navy; still one of the most formidable fighting forces in the world had been either destroyed or diverted and the Royal Air Force eliminated.

This resulted in a decisive aerial battle of attrition that became immortalised as the Battle of Britain, and it officially opened on August 13th 1940, "Eagle Day". It was one of the decisive battles of the war. Air Chief Marshall Sir Hugh Dowding, a master tactician and immensely more capable than Göring, did a first class job in resisting the demand to fling Britain's last reserves of fighter squadrons into the Battle of France, thus preserving the fighter force that met the German attempt to gain air control over Britain and the Channel. At the time, the Luftwaffe had six hundred fighters available. RAF Fighter Command had six hundred and seventy. Britain was actually out-producing Germany in fighter planes, and the proportions were steadily moving in Britain's favour.

At first German attacks were concentrated on the RAF airfields, and almost succeeded; the government issued codeword 'Cromwell', to indicate that an invasion was imminent. Church bells rang as a call to arms for the Home Guard. Across the Channel the final preparations for Operation Sealion were concentrated around their embarkation points. The two thousand five hundred transports, consisting of barges, tugs, and light craft massed in the invasion ports came under intense attacks from RAF Bomber Command and Coastal Command.

Believing that British resistance would crumble, and that the RAF would be forced to use its remaining reserve squadrons, Göring intensified the attacks and his losses mounted. These losses were shared between the fighters and bombers; whereas RAF Fighter Command, being constantly in action, bore the brunt and were soon reduced to less than one thousand pilots, who were rapidly reaching a state of physical and mental exhaustion.

However, a dramatic event intervened; Hitler had forbidden terror bombing on civilians but when a German formation got lost and jettisoned their bombs over London, Churchill ordered a reprisal raid on Berlin. Göring ordered counter raids on London thus diverting the Luftwaffe from its original purpose; that is the destruction of the RAF. This caused Operation Sealion to be postponed until September 27th, the last day for favourable tides. After that date Channel conditions would be too risky. The decision to switch objectives from British fighter bases to mass raids on London and other cities cost Germany the battle. Hitler postponed the invasion "until further notice" and ordered the dispersal of the invasion craft. Göring had failed to smash the RAF; proving that the Luftwaffe was clearly not invincible.

During the Battle of Britain, several paramount elements favoured the RAF. First was the defence radar network that although incomplete was the most technically advanced in the world. The work rate of the Hurricanes and Spitfires would have been fruitless but for this effective system of underground control centres and telephone cables, which on Dowding's initiative had been devised and built before the war. It enabled fighter planes to take off in time to avoid being attacked on the ground and directed the fighter planes by radio to intercept and often surprise the enemy.

The RAF also inflicted heavy casualties on the previously all-conquering Stukas, proving them to be most vulnerable and they were withdrawn from the battle. The British early warning system foretold any German attacks, and with the help of the code breakers of Bletchley Park, had broken the Ultra code used by the Luftwaffe. By mid-September the RAF had more pilots available than the Luftwaffe. Fighter Command had gained the upper hand and although Britain's cities were heavily bombed, by mid-October the Battle of Britain was over.

Another key element that gave Britain a massive tactical advantage was that a British pilot who survived being shot down could quickly be returned to operational status, whereas a German pilot who survived was removed from the battle and became a prisoner of war.

Hitler committed a major strategic error by allowing Göring to assume leadership in the Battle of Britain. The Reichsmarshall proved flawed in his judgment by switching air attacks from fighter airfields to London and other cities. Above all, he failed to concentrate on knocking out the radar stations.

The plan for an invasion of Britain was from the start a great risk. An unsuccessful landing would nullify all the German achievements thus far obtained and it was acknowledged that the lack of German naval and air superiority would have caused catastrophic harassment to any invasion. Hitler decided that the invasion would be executed only if there were no other ways of forcing Britain to her knees and since such circumstances were never gained, the invasion was postponed indefinitely.

Hitler diverted the German war machine to Operation Barbarossa, and was to see, as did Napoleon his great armies annihilated in the bitter Russian winter climate

Another key element that gave Britain a massive tactical advantage was that a British pilot who survived being shot down could quickly be returned to operational status, whereas a German pilot who survived was removed from the battle and became a prisoner of war.

Hitler committed a major strategic error by allowing Göring to assume leadership in the Battle of Britain. The Reichsmarshall proved flawed in his judgment by switching air attacks from fighter airfields to London and other cities. Above all, he failed to concentrate on knocking out the radar stations.

In the meantime, Hitler focused on an economic war with Britain and pursued the aim of defeating Britain in three different ways:

> *A combined air and sea attack against British trade and industry;

*Air bombardment, intended to demoralise the population,
*With the aid of his allies he would attack British positions in the Mediterranean; such as Gibraltar, Malta and the Suez Canal.

In adopting the Mediterranean strategy, Hitler quite unwittingly, began the geographical dissipation of the Wehrmacht, which in the end would prove fatal. Franco demanded too high a price for helping him take Gibraltar, and Petain was reluctant to assist in North Africa. Only Mussolini was willing, and he was an unpredictable ally.

The plan for Operation Sealion is perhaps the most flawed in the history of modern warfare. Strategic planning and preparation was woefully inadequate and would have left the German Army paralysed; its tanks standing useless without fuel and its army crippled by the lack of resources.

An explanation as to why Sealion was considered to be a huge bluff by Hitler: the German navy was in no real position to wage amphibious warfare and had no ready-made vessels suitable for landing over open beaches. Each service worked separately without a joint staff, resulting in army and navy planners soon developing conflicting ideas.

When France collapsed, in June 1940, the German staff had not even considered, never mind studied, the possibility of an invasion of Britain. Troops had received no training for seaborne and landing operations, and nothing had been done about the means of getting troops across the Channel. The Royal Navy had countless smaller craft, including sloops, minesweepers, converted trawlers and similar craft. These would have been of little value against warships. However, against the Rhine barges forming the main invasion transport force, they would have been effective.

Even if the Germans had won the Battle of Britain, a successful landing would have been a long shot. Assuming that they did establish air superiority and a beachhead in Southern England, there was still a considerable British force waiting for them, and a quick powerful counter-attack supported by the Home Fleet was a real possibility.

If the Germans did repulse this counter-attack and have a strong invading force with tanks and air cover, they now would have to capture London; one of the five largest cities in the world. It would have been held at all costs and taken months to capture, even if surrounded and besieged. Another factor is that the heavy war industries of the Midlands and Scotland would still be in British hands. Their production capacity would allow the British to continue to be supplied during any fighting. Therefore, the Germans would have to capture all of the country. Unlike France, the British would not surrender after a portion of the country was captured. The Germans would have to fight through the large urban areas of major cities such as Birmingham and Liverpool at a terrible cost in time and men.

The British could also call on huge reinforcements from India, Canada, Australia and South Africa to match the Germans in men deployed in battle. It has been suggested that an invasion immediately after Dunkirk would have produced a German success, as it would have been easier at that time. It is true that the British Army was less able to offer resistance in July than it was by September. But the difficulty facing the Germans was not beating the British Army; it was getting across the Channel in the face of the RN and the RAF.

In July the German forces had not gathered any transports and only had the capacity to lift less than one infantry division. It should be remembered that Britain had retained twenty four fighter squadrons as Home Reserve. These squadrons were rested, maintained and ready. The Luftwaffe, on the other hand, had flown many sorties in the French campaign and needed time to recover. Plus the British Radar chain was undamaged, as was the command and control structure. So the RAF was at peak efficiency in July whilst the Luftwaffe had tired crews and aircraft in need of repair.

Operation Sealion can only be described as a blueprint for a German disaster. The first steps to prepare for an invasion were taken only after the French capitulated and no definite date could be fixed. It all depended on the time required to provide the shipping, and alter them to carry tanks, and to train the troops in embarkation and disembarkation. The German invasion of Crete a year later provides an indication of what may have occurred at Sealion. Reinforcement and supply by sea proved impossible even though the Luftwaffe had absolute air superiority. The Royal Navy intercepted and utterly destroyed the flotilla of small boats crossing from Greece. Although they eventually prevailed, German paratroops and transport aircraft were decimated in the process.

One can imagine the slaughter had the RN and RAF run through the barges loaded with men and equipment during the proposed Sealion crossing. Planning an invasion and assembling a fleet in a few weeks was clearly impractical, but timing was an essential part of the game of bluff that Hitler was playing. Also, the extraordinary timing that he imposed, suggests the political rather than the military nature of the invasion. Germany did not have the industrial capacity to build specially designed landing craft for amphibious operations.

The first instruction to begin planning for Sealion was issued eighty four days before the proposed invasion date. In 1944 D-Day had been in the planning phase for two years. The parallels between Operation Sealion and Operation Overlord are striking. In every category Allied preparation for Overlord was far superior to German efforts in Sealion. On D-Day the largest amphibious force ever assembled prepared to breach Hitler's vaunted Atlantic Wall to liberate Europe.

Getting soldiers into landing craft and onto the proper beach on time is no mean feat. Plus coordinating Naval Gunfire and Close Air Support adds another degree of difficulty. The multitude of organisational and logistic considerations involved in amphibious operations is staggering. Every function in the overall plan is interdependent, relying upon precise execution for success. Most importantly, every aspect of the landing plan was reinforced with realistic training. When the Allied forces went into combat on 6th June 1944 they were physically and mentally well prepared.

Even if Fighter Command had been wiped out, RAF Bomber Command was largely intact and would have attacked the beachheads day and night. The Germans lacked the means to keep the beachheads adequately supplied and had no plans for artificial harbours or pipelines across the Channel, both of which played a crucial part in supplying Allied operations in Normandy.

As their intelligence was very poor, the Germans had little knowledge as to which beaches were the most heavily defended or the proximity of British reserves to the beaches. British counter-intelligence had already captured or "turned in" all German agents operating in Britain, which limited Germany to aerial photographs. The overall concept and execution of Overlord was a masterpiece of strategic planning made possible by the enormous capacity of Allied industry. In Normandy, the Allies had complete naval and air superiority. They also had a host of special equipment, coupled with hard-won experience, and a considerable level of support from the local population. The creation of an 'Atlantic Wall' stretching from Spain to Norway, covering some four thousand five hundred kilometers was one of the largest construction projects in human history, but, as Frederick the Great noted, "He, who defends everything, defends nothing".

Another reason why Sealion is considered a huge bluff is that at the meeting Hitler called to discuss various options, the Luftwaffe did not attend; even though it was recognised that the Luftwaffe was essential to win air supremacy and to keep the Royal Navy out of the way. The concept for getting nine divisions across the Channel was to block the west of the Channel with U-Boats, and the east of the Channel with mines and torpedo boats. The proposed time between the first landing and the second wave of reinforcements and supplies would be ten days. Thus nine attacking divisions, without any heavy equipment, would be expected to hold out against twenty nine defending divisions for this period. To get the first wave across, the Germans gathered one hundred and seventy cargo ships, one thousand three hundred barges, and five hundred tugs. The barges were mainly those designed for use on the Rhine; waves from a fast-moving destroyer would swamp and sink them.

Thus, if Royal Navy Destroyers could get close to the invasion fleet they could actually sink the lot without firing a shot. These same barges were also underpowered for open water operations, and required towing by a tug at a speed of three knots, in the Channel, which has tides of five knots. German troops would be wallowing for a minimum of twelve hours in an open boat, and then be expected to carry out a fiercely opposed amphibious landing. If this seems to be a nightmare scenario, and a recipe for disaster, it is nothing compared to other elements. The most ridiculous of which was the plan for manoeuvering the invasion barges on the landing beaches. This huge mass of towed barges was to advance in line at night coordinated by loud hailer's.

Only one training exercise was conducted off Boulogne. It was in good weather and good visibility, with no navigation hazards or enemy defences to contend with. Of fifty vessels committed, less than half managed to land their troops at H Hour. One tug lost its tow; one barge overturned when too many soldiers crowded on one side and several barges landed broad-side and unable to lower their ramps. The results of the fifty-barge exercise did not bode well for an assault on Britain.

Then there was the Irish Question. Operation Green was the German code name for the decoy invasion of Ireland, planned in conjunction with Operation Sealion in 1940. Barges were to be sent towards the south coast of Ireland to give the impression of a wide scale sea invasion of the British Isles. German agents were parachuted into Ireland to make contact with the IRA and to initiate a bombing campaign throughout Ireland to destabilise the country.

Once the I.R.A. bombing campaign got underway, German paratroops would be parachuted into zones to sever communication lines and capture RAF airfields. The Luftwaffe would then be able to strike at targets in Scotland and the west coast of England and strangle Britain's lifeline in the Atlantic. The agents, however, reported back to Germany that the I.R.A. were "unreliable" and "undisciplined" and would take months to train. The operation was finally scrapped when Sealion was placed on indefinite hold; and the German agents were later captured.

As noted previously each service vied with the other for Hitler's favour. As a result, command relationships between the services were often strained and operations suffered accordingly. Just to make matters worse, no engineers or equipment were included in the first wave to deal with obstacles. The invaders would have to cross rivers and canals more than twenty metres wide and had no means of getting across.

Then there is the question of life jackets. Thousands had been provided, but, despite all the best efforts of the planners, there were only sufficient for the first wave. According to the plan, these life jackets would be brought back again by the boats needed for the second wave. The problem was that these life jackets were worn beneath the combat pack. The troops were expected, while under fire, to first take off their pack, then their life jacket, and then don the combat pack again, and only then start doing something about those rather inconsiderate British shooting at them. One wonders what the veterans of Omaha beach would say about the viability of this. Not that it would have been of the slightest use because no one had been made responsible for collecting the life jackets and return them to the boats. The life jackets would simply have piled up on the beach.

Then there was the matter of artificial fog. A serious conflict arose between the Army and the Navy regarding the use of artificial fog. The Army wanted it for protection on the open beaches. The Navy was opposed to its use for the reason that the landings were difficult enough without making it impossible to see anything. Inevitably, a compromise solution was found; it was ruled that the Army would decide whether or not to deploy artificial fog, but that it was the responsibility of the Navy to actually deploy it.

The Luftwaffe was expected to do all of the following:

*Act as artillery for the landing forces.
*Keep the Royal Navy out of the Channel.
*Win total air superiority.
*Prevent British Army reinforcements from getting to the beachhead by bombing railway lines.
*Make mass attacks on London to force the population to flee the city and choke the surrounding roads.

With a limited range, the fighters would have a huge number of areas to protect. Meanwhile the RAF would be presented with many targets, such as barges, landing beaches and transport aircraft. If the Germans are flying fighter cover over the barges; then these fighters are not escorting German bombers, leaving them unprotected against RAF fighters. In this case, the Luftwaffe would be ineffective at keeping the Royal Navy at bay. The British came up with a far superior defensive plan such as pre-emptive attacks on staging areas, interdiction at sea and all-out assaults at the landing points. In a short period the Army refitted the survivors of Dunkirk, organised a Home Guard, created beach defences, and set up mobile reserves.

Hitler's far reaching decision to stop the Panzer's before they could deliver the coup de grace to the BEF at Dunkirk has never been satisfactorily explained. Perhaps he did not seek the outright submission of Britain. However, what is for certain is that he genuinely believed that in the end, Britain would come to terms; thereby, facilitating a rapid victory in the east that would shatter Britain's last hope of containing Germany. An impatient Hitler repeated Napoleon's mistake by trying to crush Russia before he had settled with Britain, resulting in a nightmarish war on two fronts.

Hitler failed to learn the lessons of Napoleon. He fully expected Britain to be content with a simple balance between a land power and a sea power. But as had been her foreign policy for five hundred years, Britain would only accept a balance of power on the Continent itself. Hence, Churchill's announcement to Stalin that German hegemony in Europe was as dangerous to the Soviet Union as it was to Great Britain and urged that both countries should agree on the re-establishment of the European balance of power. Although Churchill often spoke of restoring freedom to the nations of Europe, it was the balance of power that really concerned him. Something Hitler failed to grasp as basic British traditional foreign policy.

The invasion of Britain was the obvious strategic direction for Germany to pursue, but the planning was halfhearted when compared to that for the German invasion of the Soviet Union. Planning and preparation for both possibilities continued into the summer of 1941, but it was obvious by then that the campaign against the Soviets was taking shape while that against the UK was not.

Germany was not ready to do battle with a naval power such as the Royal Navy. Preparing to invade in 1941 was probably more feasible but would have required the Third Reich to focus on the production of naval and air forces at the expense of its army. Even then, if the Nazis had conquered Britain, it would not have improved their overall strategic situation. To begin with, a substantial German naval presence would have to be brought into the Mediterranean. At best, Germany would conquer Egypt, or possibly all of the Middle East by the end of 1941, turning the Mediterranean into a German lake.

However, the infrastructure to benefit from any resources found there would have to be created. Meantime, this would leave the Third Reich dependent on the Soviets for food and raw material. The Soviets could make more territorial demands on Romania; a primary source of oil for Nazi Germany. Worse yet, from Hitler's perspective, the Soviets could take advantage by attacking Germany directly. Given the huge Soviet army deployed on the eastern border of the Reich, these were not trivial concerns.

During a meeting in Berlin in November 1940, Hitler actually suggested to Molotov that the Soviets join the Tripartite Pact, despite the fact that German military planning for an attack on the Soviet Union was underway at this time. Molotov refused the offer to join the Axis, the ideological divide between the communist USSR and the fascist Third Reich was too great. Joining the Tripartite Pact entailed additional security risks for the Soviet Union; it could only benefit Germany. The Soviet refusal to join forces against Britain ensured Hitler's preference for Operation Barbarossa.

Barbarossa was preferable to Sealion for operational reasons. Landing an army across the English Channel was unlike anything the German army had ever attempted while Barbarossa required rather straightforward military planning from the German perspective. Barbarossa was a campaign that the German High Command felt competent to plan. Sealion was not. In summation, although the fear of a German invasion was real, it was unfounded. German plans were amateurish, or at best unprofessional, regarding the Channel as a relatively minor obstacle, little more than a wide river crossing. It should also be taken into consideration that Hitler's assessment of the political and military situation in 1940 was not too far wrong. America wanted no part in another world war; the U.S. Joint Chiefs, in particular, distrusted their British counterparts; they were much more focused on the growing Japanese threat.

The American Ambassador to Britain at the time was Joe Kennedy, a German sympathiser who predicted a British defeat. It was only after British survival in the Battle of Britain that America looked to lend that "fire hose" across the Atlantic and be "a good neighbor." British wartime cabinet documents released in 1998 reveals that after Dunkirk, and observing the daunting military achievements made by Nazi Germany, two members of the British War Cabinet, Neville Chamberlain and Lord Halifax, declared that they thought it best to propose a peace settlement.

There were a lot of influential British fatalists around Churchill, including Halifax, who thought the preservation of the Empire was worth an accommodation with Germany. Churchill was a relatively lone voice calling for resistance. If the RAF had not prevented the Luftwaffe from gaining air supremacy, Churchill would have been replaced by Lord Halifax, who had supported appeasement and was known to favour peace negotiations rather than face a civilian bloodbath on British soil. Churchill had to drum up support and he did this with his famous speech, on June 4th 1940, where he declared we would, "fight them on the beaches, the landing grounds, in the fields and the streets; in the hills… we shall never surrender"! The response to this was a rapturous applause in the House of Commons, and reinforced the bulldog spirit of the British people.

Although it did not appear so at the time Sealion was never a viable military option. It was a political threat that might have brought a timorous leader like Chamberlain to the negotiating table but never a tenacious warrior such as Churchill. He was keenly aware that Hitler would have to break the people on the British Island or lose the war. Further, if the Island people had been broken what would have been the consequences for the world? The British army on the Nile would be cut off without supplies permitting the Italians to dominate the Mediterranean and Suez region. The necessity to send Rommel and the Afrika Corps to Mussolini's rescue would not have arisen. Operation Barbarossa would have proceeded on schedule and there would have been no need to build the Atlantic Wall and to garrison military personnel desperately needed in Russia. The failure to conquer Britain before Russia, did not involve Germany on a war on just two fronts, but rather a war on half a dozen.

Operation Sealion was not a military plan, it was a wish.

~SIX~
THE MIGHTY HOOD

If any single ship could be said to have been the embodiment of the British Empire, and British sea power, it was the "Mighty" Hood. For more than twenty years HMS Hood upheld the pride and the traditions of the Royal Navy in every corner of the globe. By the late 1930's however, the Hood, though still a majestic warship, had begun to show her age. She was an old lady now, one of the oldest capital ships in the navy and no longer a match for more modern battleships.

On 24th May 1941, at the Battle of the Denmark Straits, the Royal Navy ships HMS Hood and HMS Prince of Wales engaged the German ships Bismarck and Prinz Eugen. The Hood was defeated and destroyed with a large loss of life. Out of a complement of one thousand four hundred and nineteen officers and men, only three were rescued. The sentimental favourite of the Royal Navy that had been looked upon with a blend of affection, admiration and awe, not only at home but the world over was gone after only a few minutes of action. Why was this, why was the 'Mighty' Hood sunk so suddenly and dramatically?

The reality of the situation is that the term "Mighty Hood" was actually a misnomer. She was not a battle ship but a battle cruiser and the difference was quite significant for the following reason. At the beginning of the 20th century, Admiral Fisher, Britain's premiere naval visionary made it his objective to ensure that the Royal Navy was not only the largest in the world but also the first truly modern navy of the 20th century. His most notable achievement was the revolutionary Dreadnought battleship that combined large calibre long-range guns with heavy armour protection.

At a stroke this innovative breed of vessel rendered all pre Dreadnought battleships obsolete and the design soon became the benchmark for all future battleships. As a consequence of this inspired leap of faith or paradigm shift, no nation or even alliance of nations could successfully stand "toe-to-toe" with the Royal Navy in fleet engagements.

Fisher came to the conclusion that the only course of action for potential enemies would be commerce raiding to strike at the empire's supply lines. The problem therefore, was how best to counter such threats and protect shipping lanes stretched around the globe? The submarine, whilst showing great potential, which the Germans would take maximum advantage, was at this time still largely unproven. Destroyers and cruisers, though fast, lacked sufficient firepower. Something new was required and this motivated Fisher to devise the "battle cruiser." He envisioned ships similar in size and firepower to the dreadnoughts, but much faster due to reduced armour protection. They would be the "cavalry of the sea" and ensure Britain's supremacy. Fisher's belief was that "speed coupled with heavy firepower would be the battle cruisers best protection". They would have the capacity to either out-run or out-gun existing warships.

Hence the concept of the "battle cruiser" was born and by the outbreak of the First World War in 1914, Britain had built a number of them. Their tactical role was, however, converted to scouting for the Grand Fleet when deployed during wartime and Fisher's argument that speed was armour would prove to be a tragic error.

On 31st May 1916, the great naval battle at Jutland took place and although the Royal Navy achieved a strategic victory, they were made to pay a high price. Three battle cruisers, The Invincible, The Queen Mary, and The Indefatigable all blew up incurring an appalling loss of life. Plunging German shells easily penetrated their deck armour to detonate magazines, thus revealing critical faults in battle cruiser design.

These tragedies made it abundantly clear that battle cruiser deck armour protection was inadequate. However, the Admiralty in general terms chose to disregard the main lessons of Jutland and proceeded to prepare designs for the new and larger "Admiral" class of battle cruisers. Therefore, future battle cruisers were destined to suffer from the same flaws as their predecessors.

Battle cruiser Hood was the largest of these new design vessels to be built for the Royal Navy. She was eight hundred and sixty feet long, had a displacement of thirty six thousand tons and was capable of reaching a speed of 32 knots. Although she was designed for speed, her 15-inch broadside firepower could match that of anything then afloat. Hood was the ultimate battle cruiser. But she had that one great defect; a lack of armour on her upper decks and events would prove that this fatal error would be the "Mighty Hood's" Achilles' heel. The design and protection of this ship were such that penetration of the magazines by high velocity armour piercing shells was quite possible. The Hood's condition could be likened to all intents and purposes as a microcosm of the British Empire at that time; ancient, glorious, biggest in the world, shrouded in ceremony, but hopelessly out of date and defective. The myth of the pre war British Empire and The Mighty Hood's invincibility was a gigantic bluff.

HMS Hood was built at John Brown shipyard in Glasgow and was named in honour of the Hood family who had given the Royal Navy four famous admirals since the 1700's. Lady Hood, the widow of Admiral Horace Hood who had been killed at Jutland when the battle cruiser Invincible blew up, launched her. The Hood possessed eight 15-inch guns mounted in pairs of four turrets; she was the fastest, the most elegant as well as the biggest warship in the world. Between the wars, when the colour red for Britain's empire covered a quarter of the globe, the Hood showed the flag on cruises to Scandinavia, South America, the Mediterranean, and the Pacific.

Her most famous cruise took place from November 1923 to September 1924, when in company with the Repulse, (another fated battle cruiser) she embarked on the "Empire Cruise". On this cruise HMS Hood logged over thirty eight thousand miles and visited numerous countries. Literally millions of people came to see her and in South Africa she called on Cape Town, East London and Durban. It was a highly successful public relations exercise and served as a subtle reminder to friend and foe alike that Britannia ostensibly still ruled the waves.

Towards the end of the 1930's when the war clouds were gathering in Europe the Hood was scheduled for a long overdue refit, which would include the reinforcement of her deck armour. This upgrade was intended to make her as potent a weapon as the newest battleships. But the threat of impending war once more with Germany caused the refit to be postponed. When The Second World War eventually broke out in September 1939, the Admiralty immediately rushed the Hood into wartime service even though she was no match for modern battleships, so why then did the Admiralty press her into service at the start of the War?

The probable answer is the "larger than life" legend and reputation that had made her so widely respected and feared the world over. This inflated reputation may also have impaired the Admiralty's judgment. She had come to be thought of as a fast battleship and had been looked upon as the "Mighty Hood" for so long that many at the Admiralty actually thought her to be invincible. In the end she proved to be, just as invincible as HMS Invincible, the very first battle cruiser to be sunk at Jutland.

In May 1940, Hitler's blitzkrieg in the west, followed by the evacuation at Dunkirk created a chaotic situation for the Anglo French allies. On June 16th, the French government, who by this stage consisted of a group of defeatists headed by Petain, informed Britain that they intended to seek an armistice with Germany. This immediately raised concerns regarding the future of the quite significant French fleet. Churchill announced that Britain would continue the fight and thereby insisted that French warships be dispatched immediately to British ports or be neutralised. The French declared that their Fleet would never be surrendered and would be scuttled if any attempt were made to seize the ships by Germany or Italy. Churchill considered this to be too risky and decided that direct action was necessary to prevent the French fleet from falling into enemy hands, which would tilt the balance of naval power in their favour.

Surprisingly the Americans who would not enter the war for another eighteen months gave their support. At Gibraltar, Vice-Admiral Somerville, the commander of Force "H" was ordered to secure the transfer of the French warships at Oran in North Africa, otherwise put them out of action. The French squadron consisted of two battleships, two battle cruisers, twelve light cruisers and destroyers, and four submarines. Force "H" included the capital ships Hood, Valiant, Resolution, and the new aircraft carrier Ark Royal.

On arrival at the French port, Somerville informed the French that their ships would not be allowed to leave harbour unless British terms were accepted. The stubborn French naval commander refused to comply with any of the conditions laid down and ordered all the French ships to prepare to put to sea. At this point Somerville, aboard his flagship HMS Hood, received an Admiralty message instructing him to settle matters quickly or he would have French reinforcements from Toulon and Algiers to deal with. The British battleships then opened fire, the first shots fired by the British against the French since Waterloo. The 15-inch salvos from Hood and other Royal Navy's capital ships overwhelmed the French vessels and the battleship "Bretagne" blew up. Shore batteries and the French warships immediately replied and heavy shells began falling near the Hood inflicting minor damage.

The French guns were soon put out of action and extensive damage was observed to a number of their ships. Force "H' then returned to Gibraltar and the Admiralty expressed deep regret to France for the tragedy in which almost one thousand four hundred French sailors had been killed. The Vichy government was outraged and ordered immediate reprisals against their former ally. Consequently diplomatic relations with Britain were severed and French planes attacked Gibraltar but little damage was caused. France actually came very close to declaring war on Britain and bitterness lingers amongst the French to this day regarding this incident.

The next major action involving the Hood Almost a year later, proved to be her last. In 1941 Germany brought the Bismarck into service. This magnificent battleship represented the state of the art in German ship design and was the most powerful warship afloat. Grand Admiral Erich Raeder, head of the German navy was hopeful that the surface warships combined with the U-boats would successfully blockade and starve Britain into submission. Britain was barely surviving at this stage of the war.

Bismarck was of comparable size and main armament to HMS Hood, but here the similarity ended. The German ship was far superior to the Hood, possessing excellent up-to-date electronics and superb armour protection. In effect, Bismarck was twenty years in advance of the Hood and this was reflected by the high quality technological developments which had taken place since the Hood was built.

In April 1941 the German navy planned to form a battle group combining Bismarck, Prinz Eugen, Scharnhorst, and Gneisenau, for Operation Rheinübung. This lethal squadron was the nucleus of the big battle fleet then being developed and would be capable of taking on even the most heavily defended convoys and would have the necessary speed to escape. These fast powerful vessels in the great spaces of the Atlantic Ocean would subject the Royal Navy to a trial of the first magnitude. Churchill was aware of the danger this threat posed and ordered the RAF and Coastal Command to make every effort to sink them regardless of the risks involved.

At this stage Scharnhorst was forced to withdraw due to troublesome boilers; thus reducing the operational strength of Rheinübung to Bismarck, Gneisenau and Prinz Eugen. On 5th April 1941 during an RAF raid on Brest where the Scharnhorst and Gneisenau were based, an unexploded bomb landed in the dry dock where Gneisenau was berthed. The German battle cruiser was moved out to a mooring in the harbour. The next day Flying Officer Kenneth Campbell's Beaufort torpedo bomber attacked her. The Gneisenau was hit, a propeller shaft was destroyed, and two engine rooms flooded. This put the Gneisenau out of action for six months.

Due to the crucial action of Kenneth Campbell VC, Operation Rheinübung was now reduced to Bismarck and Prinz Eugen.

Unwilling to delay the operation, Grand Admiral Raeder insisted on sending Bismarck into the Atlantic convoy routes accompanied only by the new heavy cruiser Prinz Eugen. Raeder was influenced by his First World War experience where the idle German fleet became a breeding ground for communist revolutionaries that led to the mutiny at the end of WW1.

On 18th May 1941, only six weeks after Campbell had crippled the Gneisenau, the two warships sailed from Germany bound for Norwegian waters. This would be Bismarck's first and last operation. The primary objective once Bismarck and Prinz Eugen_had broken out into the Atlantic was not to engage in fleet actions but to attack British merchant convoys; the Bismarck to deal with the escorts and the Prinz Eugen to dispatch the merchantmen. Together with the U boats they were intended to present a menace the like of which the British had never had to contend with before. At this time, eleven convoys including a troop convoy carrying twenty thousand men destined for the Middle East were at sea. Also, from an overall strategic point of view it was desirous for the Germans to draw the Royal Navy away from Rommel's supply lines in the Mediterranean and at the same time prevent British naval interference in the invasion of Crete. In this they were successful; Churchill forbade the release of Force "H" from Gibraltar to support the defence of Crete. This was now out of the question due to the threat of Bismarck.

The British Home Fleet Commander-in-Chief Admiral Tovey was fully aware that both Bismarck and Prinz Eugen had completed their training and were about to put to sea. He had to make his dispositions as a matter of urgency to meet this impending menace. He decided, however to remain at Scapa on his flagship King George V until the situation became clearer. Also anchored at Scapa Flow under the command of Vice Admiral Holland were the new battleship Prince of Wales and battlecruiser Hood. Tovey now evaluated the various scenarios posed by the emergence of Bismarck and Prinz Eugen.

The three obvious possibilities he had to consider were;

*They may be merely escorting a troop convoy to northern Norway.
*They may be part of a raiding force bound for Iceland
*They may be attempting to break out into the Atlantic to attack convoys.

Tovey thought the third possibility was the most likely and presented the greatest risk. Once out they would wreak havoc on the convoys and be almost impossible to track down. They had to be stopped before they achieved their Atlantic objective.

Therefore, assuming the Atlantic was their objective there were three corridors for entry;

*Between the Shetland Islands and the Faeroes.
*Between the Faeroes and Iceland.
*The Denmark Strait north of Iceland.

The first was dangerously close to Scapa Flow. The second was permanently patrolled by British cruisers, and the third was very narrow owing to the ice pack and a British minefield. On this basis, Admiral Tovey had no alternative other than to divide his naval forces to cover all possible contingencies and to have strike force capability at all of these routes. The radar equipped British cruisers Norfolk and Suffolk were ordered to patrol the seas between Greenland and Iceland. Other cruisers were also on patrol but we will focus on Hood's action. Aboard the Hood, Vice-Admiral Holland received orders that together with the Prince of Wales and screening destroyers they were to sail towards the Denmark Straits.

The brand-new Prince of Wales sailed to intercept Bismarck with civilian workers still on board making adjustments to her main armament. Prince of Wales had only recently been commissioned and was not yet fully worked up. On the morning of 21st May Bismarck accompanied with the heavy cruiser Prinz Eugen, entered German occupied Norwegian waters. Their movements did not go unnoticed; an RAF reconnaissance Spitfire spotted the warships at anchor and a Swedish cruiser had also reported her to be at sea. The British Admiralty now knew exactly were Bismarck was but what her intentions were; was another matter.

On 22nd May, both ships sailed from Norway. But it was still uncertain which route they would take on their way into the Atlantic. Air reconnaissance now revealed that they had left Norway. Tovey sailed from Scapa Flow on board the King George V, accompanied with the aircraft carrier Victorious. On 23rd May, they were joined by the Repulse that had sailed north from the Clyde. The entire force set a course to patrol an area covering the Iceland Faeroes passage. The foggy weather provided an observation shield for the German warships as they entered the Denmark Straits. It had been twenty nine hours since Bismarck had last been sighted in Norway. If the fog held, they might be lucky and slip through. Their speed was increased to 27 knots in an attempt to ensure that the breakout was achieved before the weather cleared.

But, unfortunately for the Germans, the conditions changed. In the late afternoon of the 23rd the fog lifted and visibility improved sufficiently to enable HMS Suffolk to sight the Bismarck. The cruiser radioed reports of her position and course. Hood and Prince of Wales were then about three hundred miles distant. Suffolk manoeuvered to place herself astern of the enemy ships and began shadowing and monitoring them with her radar. Bismarck also spotted Suffolk and immediately opened fire; it was the first time that the new German battleship had fired salvos in anger. Admiral Holland turned to close on the enemy and ordered his ships to increase speed to twenty-seven knots and to steer on an interception course, which was expected to take place the next morning. The crews aboard both British ships were called to action stations and raised battle ensigns. Admiral Holland then decided on his battle plans. The first option was to cut across their bows with the classic "T". This would allow all the British guns to bear on the German ships whilst the enemy would only be able to fire with their forward guns.

The second choice was to cross well ahead, then swing around and approach from the west. This would silhouette the Germans against the morning sky and considerably ease range finding for the British.

Either of these plans might have worked except that at this stage Suffolk lost contact with the enemy. This disastrous news reached Vice-Admiral Holland on the Hood shortly after midnight when they were an estimated one hundred and twenty miles south of the German ships. With no definite position or bearing for the German warships, Holland ordered that the crews go to "relaxed action stations" and reduced speed to 25 knots. For Admiral Holland the pursuit had gone entirely wrong. To enable him to decide on his best course of action he required constant and accurate reports from Suffolk on the enemy's precise position. He was now "searching in the dark" and it had become a guessing game.

Loss of contact by Suffolk was a devastating blow. He now had to consider the possibility that the Germans may well have altered course. This being the case, Holland had to decide what that course alteration might be. The Germans may have decided that having been located and tracked, British forces were likely concentrating to intercept them and it would be better to reverse course and disappear into the Arctic Ocean. Alternatively, they may have changed course to the southeast or the southwest. In any case, Holland decided to close the distance between his ships and the last known position of the German squadron as quickly as possible. He also had to maintain radio silence so that Bismarck would not be alerted of his presence.

Just before 03:00 hours Suffolk signalled that she had regained radar contact with Bismarck and Prinz Eugen. Holland altered course to converge with the enemy and speed was increased to twenty-eight knots. Suffolk's reports placed the enemy approximately thirty five miles northwest of Hood and Prince of Wales. They had indeed altered their course and Holland was now in a dangerous and exposed position. Instead of closing the Bismarck to cross her "T" or attack under the cover of darkness, he was closing head on which gave the enemy the advantage. Holland ordered, "Prepare for instant action." The crews then went to the first level of readiness. He expected Norfolk and Suffolk to engage Prinz Eugen, but because of the need for radio silence this was not confirmed.

The command crew trained their binoculars and strained their eyes to the north, as they silently waited for contact to be made. Over the past few hours visibility had gradually increased and at 05 35 hours, observers in Prince of Wales, followed almost immediately by those in Hood, visually sighted the enemy vessels at a range of about seventeen miles. Over seven thousand officers and men on four great ships were racing towards a fateful meeting.

The Germans were well aware of the approach of the British warships. Hydrophones aboard Prinz Eugen had detected the sounds of fast moving turbine-driven vessels some time earlier and assumed them to be enemy cruisers. This was because their intelligence had informed them that all the British capital ships were still at Scapa Flow.

As the vessels drew ever closer, observers on Prinz Eugen and Bismarck sighted the British vessels. They were shocked to see that the growing silhouettes were in fact a modern battleship and even worse, accompanying it was the famed and feared Mighty Hood. This eventuality placed Admiral Lütjens aboard the Bismarck in somewhat of a quandary for he had strict orders not to engage enemy capital ships unless absolutely necessary. His main priority was to get to the Atlantic unscathed. But he had no choice; the Hood was one of the few ships that had sufficient speed and firepower to match the Bismarck.

Admiral Holland aboard the Hood, decided to make a head-on dash at the enemy, and then turn at short range to bring his full guns to bear. His original battle plans had gone awry and he knew full well of Hood's susceptibility to long-range plunging shells. It was therefore critical that Hood close the range as quickly as possible. Once the range had been reduced, Hood would of course, still be susceptible to enemy shellfire, but at shorter ranges their broadsides would come at flatter trajectories. Hood's main side armour during such an engagement should stand up nearly as well as Bismarck's or Prince of Wales'.

However, there were severe risks and disadvantages to the approach he had adopted; in that he presented the Germans with the advantage of bringing all their guns to bear. The Hood would only be able to bring half of her eight 15 inch main guns to bear as opposed to Bismarck's eight 15 inch guns. The substantial advantage was lost. In the first all-important minutes of battle the relative weight of broadsides was overwhelmingly in the enemy's favour. Holland also had his ships in close formation, Hood in the lead with Prince of Wales only one thousand yards off her starboard. This meant that when the enemy found the range, it would be fairly easy to shift fire from one ship to the other. Tactical superiority now passed to the Germans; Holland was later to be criticised for going into battle with "one hand tied behind his back".

At this point Admiral Tovey aboard the King George V considered signalling to Holland to station the Prince of Wales ahead of The Hood so that the better-protected ship might draw the enemy's fire. But he decided against it, not wishing to interfere with a senior commander's operation. He was very soon to profoundly regret that he had not made the signal. Admiral Holland ordered fire to be concentrated on the left hand ship, which had been erroneously identified as Bismarck. In fact, the leading ship was Prinz Eugen, and the gunnery officer of Prince of Wales, realising this, transferred his attention to the second ship in the enemy line.

The mistake was also realised on the Hood. Shortly after her opening salvos, she hoisted the signal "shift target right". The reason for the misidentification was that the German vessels had switched positions the day before when Bismarck's forward radar had been put out of action when she had fired at the Suffolk. Hence the switching of positions, Bismarck did not wish to lead blind. Also, given the extreme ranges and angles at which the enemy ships were first sighted, they were virtually indistinguishable as they both had very similar silhouettes.

Holland proceeded to make several more mistakes. His first, as previously mentioned was to throw away his superiority in firepower. His second was to advance in close order, and his third was to incorrectly identify the leading enemy ship. On top of this why did he not use his destroyers to launch torpedo attacks during the gun battle and why did he not break radio silence after the enemy was sighted and allow his cruisers to engage Prinz Eugen?

The Germans, on sighting their adversaries, changed back to their original positions; Bismarck once more taking the lead with Prinz Eugen astern. Bismarck now trained its secondary armament on

Prince of Wales, but kept the main guns fixed on Hood. The full weight of eight gun broadsides from both German ships was about to be brought to bear on Hood. Action commenced at 05 52 hours. Hood's two forward turrets opened fire at a range of twenty five thousand yards. Moments later, Prince of Wales' forward turrets followed suit. Hood was able to use her gunnery radar but Prince of Wales' radar was malfunctioning.

The enemy opened fire on Hood. Bismarck's first salvo fell short, the second was over and the third straddled. The 15-inch shells weighing a ton apiece went rocketing out of the muzzles at over one thousand six hundred miles an hour and the men on either side anxiously counted the seconds until their arrival. Holland having reduced his initial superiority in heavy guns from eighteen against eight to ten against eight, found this further reduced to nine against eight when one of Prince of Wales's forward guns developed a defect. Both German ships continued to concentrate their fire on Hood. Prinz Eugen's fire was the most accurate. She scored hits with her 8 inch shells that immediately enveloped the Hood to start explosions amongst the anti aircraft ammunition. The bright fire burning on the Hood as a result of hits by Prinz Eugen gave the Germans a much better aiming point with their stereoscopic optical range finders.

The fire made it much easier for them to determine accurate range and bearing information. It would have helped if Norfolk and Suffolk had closed up and worried the German ships from the rear to draw the fire from her after turrets, which was Holland's initial intention, but had failed to communicate and the opportunity was lost. Another handicap for Holland was steering into the wind, which caused spray to drench the lenses of the British range finders. Vice Admiral Holland must have realised that the situation was getting desperate - the Germans had already found the range and Hood was taking severe punishment causing terrible casualties.

If Admiral Holland's original battle plan had been put into effect, and he had come upon the enemy ships unseen, he might already have won a great victory. But now everything had gone sour on him and the enemy shellfire was deadly accurate. By steering at this angle he was still denying his own force the maximum of firepower. Worst of all, neither of his ships was scoring any decisive hits on the enemy. Hood, having made the error of initially opening fire against the wrong ship was only now getting the correct range for Bismarck. He had to bring more guns to bear as quickly as possible, and at long last, the signal was given for the ships to turn to port in an attempt to come more broadsides on to the enemy and bring the rear turrets into action. With the range down to about fourteen thousand five hundred yards, Admiral Holland ordered his ships to turn twenty degrees to port so as to bring their full broadsides to bear. As this order was being executed, Bismarck's fifth salvo straddled Hood, scoring one or perhaps two hits with disastrous and incredible results. The end of the pride of the Royal Navy came spectacularly only eight minutes after the start of the action. A sheet of flame was seen to leap into the air from the vicinity of Hood's mainmast, followed by a tremendous explosion that broke the ship in two. There were three survivors from the crew of one thousand four hundred and nineteen officers and men.

Prince of Wales had to take immediate evasive action to avoid Hood's sinking wreckage. This resulted in her loss of aim and the emergency turn also placed Prince of Wales directly in the sights of the enemy ships. The close formation that the British ships had taken up now meant that Bismarck's gunners could quickly and accurately switch fire to the Prince of Wales. The situation had dramatically reversed. Bismarck's elated gun crew focused on the battle once more and with deadly accuracy scored several direct hits on the Prince of Wales. The first 15 inch shell to find its mark went straight through the compass platform, killing all personnel except Captain Leach, and the Navigating Officer. It also wrecked most of the control and communications instrumentation.

Four more 8 inch hits from Prinz Eugen and two more 15 inch hits from Bismarck soon followed. These included a 15 inch hit below the waterline and beneath the armoured belt. Fortunately for Prince of Wales, this shell, which was potentially fatal, failed to explode. With her fighting capacity drastically reduced she was soon in dire straits and now became the target for the concentrated fire of both enemy ships. Bismarck's salvos continued to thunder out every twenty seconds; Prinz Eugen's every ten. Captain Leach realised that if he continued much longer in this unequal battle, he might soon deprive the Royal Navy of a valuable ship without inflicting further damage on the enemy. By this time the Prince of Wales was within range of Prinz Eugen's torpedoes. He, therefore, made the decision to lay down a smoke screen and break off the engagement.

This decision was also prompted by the fact that 'A' turret had jammed and was partially flooded, 'Y' turret was out of action and three of the 14 inch guns had jammed in elevation. The ship was clearly not in fighting condition. Despite the numerous hits sustained and the ever increasing problems with her guns Prince of Wales did manage to score three hits on Bismarck - one of which was to prove critical to the outcome of Operation Rheinübung.The battle had only lasted twenty four minutes and British casualties were heavy. The Germans suffered no casualties.

Unbelievably, Hood was gone. All that remained was a morass of floating debris and a huge oil slick. Only three of her crew survived, Midshipman William Dundas (who ironically was killed after the war in a car accident), Able Seaman Robert Tilburn and Signalman Ted Briggs. There was no trace of anyone else living or dead; all had perished. One week before the twenty fifth anniversary of the Battle of Jutland she blew up just as the three battle cruisers at Jutland had. The battle had been a shocking German victory.

Falling almost vertically from above, possibly two shells are believed to have struck the twenty five year-old Hood with incredible violence. These plunged downwards through the unstrengthened deck, as straight as arrows, as if cutting through cardboard. The ship's vitals deep below the water line were hit and a gigantic explosion wracked the entire rear of the ship. Fierce upward flames resembling a volcano erupting enveloped the ship from one end to the other. Survivor, Ted Briggs later recorded that Vice Admiral Holland had sat dejectedly in his seat, with Captain Kerr standing at his side. No order was given to abandon ship; it really was not necessary, the ship was abandoning them. Everyone seemed to realise what was happening; Hood had taken a mortal blow. Those able to see the action from Prince of Wales, Norfolk and Suffolk, as well as their German adversaries, could not believe their eyes - the Mighty Hood, most famous of all warships, had just been devastated by a massive explosion. Large pieces of the ship including the turret and main mast were seen hurtling through the air.

It truly was nightmarish to all who watched. The aft portion of the ship, some three hundred to four hundred feet, had been laid waste - a mass of largely unrecognisable steel and twisted framework. The eyewitnesses watched in horror as the stern or what was left of it, rolled over to port and immediately sank. The forepart swung high into the air at an angle between 45° and vertical, pivoted about and began to sink rapidly. According to the Germans, as the bow rose into the air, Hood's forward turrets were seen to fire one last salvo. This was likely due to a short in the circuits or a mechanical failure. But to her enemies it seemed a last defiant and courageous gesture.

Bismarck continued on her southwesterly course shadowed from a respectable distance, by Norfolk, Suffolk and now the Prince of Wales. Aboard Bismarck, Admiral Lütjens was satisfied with the outcome. He was mindful that the over-riding objective of Operation Rheinübung was that Bismarck and Prinz Eugen should break out into the Atlantic. He, therefore, ignored the heated suggestions from Captain Lindemann that they pursue and destroy the damaged Prince of Wales.

He was of the opinion that any delay would probably bring other British warships to the area and considered it best to make a quick getaway. Besides, the Bismarck did not emerge from this action unscathed. She had also taken hits during the engagement and the damage needed to be assessed and repairs effected as quickly as possible.

There were three possible courses of action that Admiral Lütjens could now pursue;

*He could continue with his primary operation.
*He could go after the Prince of Wales and finish her off.
*He could bring his squadron home to a hero's welcome for the unprecedented triumph of sinking the enemy's proudest battleship.

The Bismarck had sunk the pride of the Royal Navy and given a terrible thrashing to its newest battleship. She could have returned to Germany satisfied with what amounted to a resounding triumph.

Her prestige and striking power would have been immensely enhanced. Potentially greater success could later be achieved with a combined breakout with the Gneisenau, Scharnhorst, and possibly even the Tirpitz added to the present squadron of Bismarck and Prinz Eugen.

However, the damage assessment revealed that Bismarck had been hit three times by 14 inch shells from Prince of Wales. One shell had penetrated an oil fuel tank causing a serious oil leak and contaminating the oil in adjacent tanks. Another had struck the side armour amidships, causing flooding and putting one dynamo and one boiler out of action to reduce her maximum speed by two knots. The third hit caused only minor damage. Admiral Lütjens decided to make for occupied France where repairs could be carried out. As a result, Bismarck's proposed Atlantic sortie was abandoned. Prinz Eugen, unharmed in the battle, should attempt commerce raiding on her own, whilst Bismarck was to proceed to the port of Brest.

Returning to the Hood survivors, the three men had each found their way to rafts floating in the wreckage. They then managed to link up with one another and for nearly four hours they fought off the killing effects of hypothermia. Besides being thoroughly soaked, they were caked with oil and in a state of shock. As they clung on in the icy sea, they watched the severely mauled Prince of Wales steaming past attempting to continue the battle. They could also see smoke from Norfolk quite a way off. The men eventually drifted apart and began to give up hope. Fortunately, a destroyer arrived on the scene to pick up the three men. She searched for other survivors, found none, and then departed the scene.

For the British public, the news of the Hood's death was traumatic, as though Buckingham Palace had been laid flat or the Prime Minister assassinated, so integral a part was she of Britain and her empire. Throughout the world the news had a shattering impact on all who had remembered this magnificent ship gliding gracefully into their harbours. Just as people today can tell you where they were when President Kennedy died so to in wartime do they remember the Hood's end. Many felt the war was lost, if Hood could not stop the Bismarck what could? What now lay between this mighty German battleship and the destruction of the Atlantic convoys?

Nearly every officer on duty in the Admiralty was stunned. Most of them had served on the Hood and remembered her as the pride of the Royal Navy. Now that she was gone, resolution set in to avenge her death at all costs. They, therefore, redoubled their determination and plotted Bismarck's death. Admiral Summerville's Force "H" of which Hood had recently been the flagship was ordered to sea from Gibraltar. The battleships Rodney and Ramillies broke off from convoy escorting duties, and battleship Revenge sailed from Halifax.

The Hood was finally avenged. In a truly epic naval drama, the Royal Navy found and sunk the Bismarck. The German death toll was over two thousand. Bismarck's defeat at the hands of the Royal Navy was doubly crippling to the German fleet, for after the British victory on 26th May the Bismarck's equally formidable sister-ship, Tirpitz, rarely dared to venture out of port, spending most of the war hiding in a Norwegian fjord.

Two boards of inquiry were convened on the loss of the Hood. Both inquiries concluded that the cause of the Hood's destruction was due to one or perhaps even two 15 inch shells that pierced through the thin deck armour to set off the magazine. However, not all experts agreed with the findings, causing some lingering doubts to remain to this day of the Royal Navy's largest loss of life on any ship during the Second World War.

In the final analysis, the patriotic emotion that swept the nation may be concluded as a fitting epitaph and tribute to "THE MIGHTY HOOD."

~SEVEN~
OPERATION BARBAROSSA

The critical issue of World War Two, if not of the twentieth century, may be generally regarded as Adolf Hitler's decision in 1941 to launch an assault on the Soviet Union. At this time Hitler was master of almost the entire continent of Europe and in an attempt to end the war, extended peace offers to Great Britain. However, for reasons Hitler was unable to comprehend, and despite the hopelessness of her position, Britain rejected these proposals, thus frustrating his aim of bringing the war to a successful conclusion.

Consequently, he turned his back on Britain and attacked the Soviet Union, to launch a war on two fronts, contrary to his pledge to the German people. The assault on Russia resulted in an unprecedented titanic struggle that no other theatre of operations in the Second World War would compare. It caused thirty million casualties and only ended with the complete annihilation of the Wehrmacht in the ruins of Berlin.

"Operation Barbarossa" the code name for the German invasion of Russia was planned to start in mid May 1941 and be completed before the onset of the Prussian winter. Hitler was, however, compelled to postpone the attack for a vital five weeks to intervene in the Balkans crisis due to the actions of his over ambitious Italian ally. The surprise "Blitzkrieg" attack was nevertheless confidently launched against the unprepared Soviet Union on June 22nd 1941.

The die was cast, Hitler was determined to gamble Nazi Germany and set out to conquer a country the size of the North American continent. All of the Soviet Union from Archangel to the Volga was to be drawn into the greater Reich and the new German empire. This throw of the dice was to prove catastrophic to his thousand year Reich for he would reap no easy victory in Russia. Hitler's decision is perceived as the great strategical mistake that doomed Nazi Germany to defeat, for although he willed the Germans to give their best shot, he had vastly underestimated Soviet potential.

For decades this has been accepted as the official version of the German-Soviet conflict. However, how accurate is this explanation? Many historians have often pondered on this, as it seems senseless for Germany to attack the Soviet Union, with whom they had signed a non-aggression pact, and for all intents and purposes were now allies. Mainly due to the collapse of the Soviet Union in 1991, revisionists nowadays are challenging this conventional point of view. Particularly as the new Russian government has opened up the archives of the Soviet Union to research; after all they have no reason to protect the preceding regime. Various documents claim that it was actually the Soviet Union who was preparing to launch an attack, and not only against Germany but all of Europe and this was scheduled to take place on July 6th, 1941.

Revisionists provide compelling information that the German attack was a pre-emptive strike against a massive Soviet military build-up that was poised to strike. Stalin, it is claimed schemed the European war to enable the USSR to intervene at the most expedient moment. To ignite this he used the Soviet-German Non-Aggression Pact to provoke Hitler's attack against Poland and led to the declarations of war by Britain and France. This presented an ideal opportunity to advance world socialism and extend the Soviet Empire. Red Army forces were, therefore, transferred from the interior to the western border, and deployed at take-off points for the planned offensive.

The objective was to destroy the German armed forces and with Germany and France under Soviet domination, Italy and Spain would also have quickly fallen. Stalin's one million paratroopers would then have made short work of seizing the airfields of southern England to clear the way for a full-scale invasion. The Soviets actually anticipated that the isolationist United States would accept their planned conquest of Europe. Later events would indicate that this was perhaps not unreasonable considering that the Americans, much to the consternation of Winston Churchill, ceded much of Europe to the Soviets in 1945.

But returning to the events of 1941, prior to the German attack on June 22nd, more than half the Soviet forces were deep in Polish occupied territory. German intelligence was not oblivious to these massed Soviet formations deploying on their frontier and adopted counter-measures to these developments. Therefore, Operation Barbarossa was launched as a pre-emptive strike to forestall the imminent Soviet invasion of Europe.

Since 1945 historians have debated German strategy in the campaign against the Soviet Union. Some maintain that Germany could have won the war in the summer of 1941, had they struck straight for Moscow, rather than turn south to capture Kiev. It is postulated that the Soviet Union would have capitulated if Moscow had fallen. A further issue for consideration is that when the attack was launched, German hopes of victory were optimistically built on the prospect that the invasion would produce a political upheaval in Russia.

Hitler stated beforehand that there was little chance of achieving final victory without the help of such an upheaval. Telling his generals: "We have only to kick in the door and the whole rotten structure will come tumbling down and the world will hold its breath". He believed that the Russian people, as in World War One, would revolt against the political leadership if they suffered heavy defeats. Therefore, no German preparations were made for a prolonged struggle. Everything was based on a decisive result before the winter set in. Before speculating further, it is necessary to look at what certainly did happen.

The largest and most powerful German army ever assembled opened the greatest land battle in history with a massive three pronged thrust into Soviet territory. Army Group North headed towards Leningrad. Amy Group Centre for Moscow and Army Group South the

Ukraine and Crimea. This is how the war came to Russia! And the world did indeed hold its breath, except that is for Winston Churchill who announced, “if Hitler invaded hell, I would at least make favourable reference to the devil in the house of commons".

The main German objective was to destroy the Soviet armies at the frontier before they had time to mobilise their immense reserves of manpower. The Luftwaffe rapidly gained air superiority and provided invaluable support for the ground troops. The Russians lost one thousand two hundred aircraft on day one, most of them destroyed on the ground; the Germans lost ten.

Whilst the Soviet air force was being decimated, the Red Army, suffering terrible losses, was in full retreat everywhere and German armoured formations penetrated as far as eighty kilometers on the day of invasion. Minsk fell on July 9th entrapping three Russian armies. The Germans took three hundred and twenty five thousand prisoners, three thousand vehicles, plus one thousand eight hundred field guns. Amongst the prisoners was Stalin's son.

The German High Command then became nervous about the speed of the advance and ordered the panzers to slow down to allow the infantry to catch up. Guderian, however, would have none of this and risking court martial disobeyed orders and pushed on to further distance the mobile columns from the infantry. Stalin broadcast that the Germans were not invincible, and reminding the Russian people of Napoleon's fate, ordered a scorched earth policy. The superb Russian T34 tank then made its unexpected appearance to startle the Germans and inflict heavy casualties. It outgunned and outmatched German tanks and was designed to move across the terrain in weather that immobilised the panzers. The T34 was at this stage immune to anti-tank shells that simply bounced off its thick sloping armour. Only superior tactics plus the conversion of the 88mm anti-aircraft gun to an anti tank role overcame this problem.

Smolensk fell next, and Soviet losses were similar to that as at Minsk. The Germans believed total victory was near and there seemed to be no reason why they should not reach the Soviet capital as planned. However, a significant development occurred, Hitler vacillated on the drive to Moscow, declaring that it was:- "merely a geographic expression". He ordered the transfer of Army Group Centre's resources to support Army Group South. Guderian bitterly protested at having to divert one thousand kilometers south when Moscow was only three hundred and fifty kilometers ahead, but, Hitler was adamant; the direct attack on Moscow was to be suspended. Kiev was now the German primary objective. Stalin ordered the city to be held at all costs and this proved to be a disaster. This permitted Rundstedt's double envelopment to be a brilliant tactical success and Kiev was taken. Into the bag went six hundred and fifty thousand prisoners, one thousand armoured vehicles, and four thousand field guns. It was the most comprehensive defeat ever experienced by the Red Army and perhaps any other army for that matter.

At this time armoured units of Army Group North reached the outskirts of Leningrad. Hitler ordered them to stop; Leningrad was to be starved into submission. This is claimed as a blunder to equal Dunkirk. Nine hundred days of siege was endured and half of the three million population perished.

Hitler now approved the assault on Moscow, but ominous signs were appearing that the Soviet Union would not fall as easily as predicted. Tank engines wore out quickly and thick dust clogged the air filters. Insufficient supplies were getting through to the front due to demolished Soviet railway tracks. Heavy rains then turned the roads to mud and grounded the Luftwaffe. The German juggernaut slowed down to a snail's pace. German front line units were living on a "hand-to-mouth" existence and losses were already higher than in all the previous campaigns combined. The Wehrmacht had lost over half a million men and lacked replacements. Guderian was reduced to only nine battle-ready tanks and ammunition shortages stalled the infantry.

It was now realised that instead of having to destroy the expected two hundred Soviet divisions, more than three hundred and fifty had already been identified, and the Russians had evacuated most of their plant and machinery to re- establish them in factories beyond the Ural mountains. Huge bypassed Soviet concentrations, refusing to surrender, escaped into the forests and swamps to snipe at German supply columns. This was also true of civilians, women, children and old men went into hiding and formed guerrilla units. The door had indeed been burst open but the building would not collapse. However, despite this nightmare, the German army had made enormous territorial gains. The frosts came in November thus enabling the German forces to embark on their final lunge towards Moscow. The German high command was convinced that Russian resistance was at breaking point and in a desperate effort threw in their last reserves.

The Russian armies withdrew to establish the defence of Moscow but the Germans did not give them time and tore through to encircle them. Russian losses were again staggering and there were no further combat defences before Moscow. The Soviet Union was perilously close to defeat! An elated Hitler announced – "Russia is broken and will never rise again". The Germans were now established one hundred kilometers from Moscow. Panic swept the capital, which was abandoned by many government officials and martial law proclaimed. Two million people left Moscow and headed east. Stalin decided to call in general Zhukov to take command and save the capital. By November 28th, German troops penetrated as far as Moscow's outer suburbs to observe the towers and domes of the Kremlin. That was as far as they got. .

Hitler's armies had reached the limits of their endurance. Having suffered seven hundred and fifty thousand casualties, their fighting capacity was on the verge of exhaustion. Russian tactics of gradually wearing down the German forces had served its purpose. Good fortune now smiled on the stricken Soviets. Mother Russia's powerful ally - General Winter again came to her aid. Heavy snow and sub-zero temperatures thwarted all German efforts and took its toll. In temperatures of – 40 c, machine guns froze, oil turned to sludge, batteries died and engines had to be kept running continuously to consume fuel supplies. Anti-freeze was unavailable.

Troops heated bricks on stoves and hastily placed them against guns to keep them operational. The frozen troops had no real winter clothes and their ordinary uniforms were totally inadequate. German soldiers wearing summer uniforms would pay a terrible price with their dark green uniforms silhouetted against the winter snow. Goebbels' appeal for warm clothing for the troops in Russia must have been an ominous sign to the German people, Despite the propaganda efforts, all was not well on the eastern front. The offensive staggered to a halt, the soldiers did not have the equipment to survive let alone fight. Hitler's war on the Soviet Union was not over - it had only just begun.

Stalin was convinced that countries in Western Europe had been beaten by their own fear of German invincibility and he sensed that Blitzkrieg could be defeated. He ceased to fear an assault on the Soviet Union from the Far East. Intelligence sources reported that the Japanese were intent on attacking American and Colonial possessions in the Pacific. This permitted well-equipped winter trained Siberian troops to be transferred west to counter-attack the under strength Germans on the outskirts of Moscow. Their arrival turned the scales in the battle for Moscow.

Zhukov launched a huge pincer movement with these seasoned troops in their white winter gear and thick fur lined boots, who were trained to lie out in the open for hours or even days at a time. They attacked along a one thousand kilometer front with the intention of encircling and crushing the vulnerable and exposed Wehrmacht. The over strained German supply system broke down. Units were short of food, fuel, ammunition and medical supplies; frostbite was striking down thousands of men. They had little time to prepare defensive positions in the hard ground, thus rendering the German position untenable.

Over the entire front dangerous breaches were opened up and the German armies were fragmented into isolated units. Hitler ordered his troops to stand fast, but after intense and bitter fighting, the Russians made significant advances. However, they lacked the experience of the Germans in coordinated attack after a breakthrough. Hitler took over command of the armed forces and turned his fury on his generals for daring to give rather than hold ground. Administering a great shake-up in the German command, he sacked Brauchitsch, Leeb, Bock, Rundstedt and Guderian. Over a period of just five days he removed a total of thirty five corps and divisional commanders. The former corporal now took personal charge of the war in the east.

As it turned out, Hitler's decision to stand fast rather than withdraw was actually correct under the circumstances, and saved the German army from total disintegration. The Wehrmacht although badly mauled, was spared the fate of Napoleons Grand Army. By February the Russian attack ran out of steam, allowing the Germans to stabilise and consolidate the front. However, the myth of German invincibility had been shattered. The Soviet Union had not collapsed and the Red Army had repulsed their aggressors.

Despite exceptional tactical victories and colossal Soviet defeats, Hitler never gained his principal objectives:- Leningrad, Moscow or the Crimea.

Hitler had started too late and attempted too much and now in defeat blamed everyone except himself. So, what went wrong in Russia, considering that in 1941 the Wehrmacht was the most highly trained and battle seasoned fighting force with superior skills at all levels. The German troops had an unbroken string of victories to their credit and the German general staff was considered the best in the world. The remarkable achievements of the Germans are undeniable. The Wehrmacht had carried out the greatest sustained offensive in military history. But, German manpower and equipment ultimately proved to be inadequate in the vastness of a Russian hinterland that could not be conquered, even though the Soviets had at the time of the invasion troops that lacked good leadership, tactics, and more importantly, experienced officers. Stalin's purges and outdated tactical know-how was a key factor in the opening days of Barbarossa.

German initial successes may be attributed to tactical preparations, surprise, and the Luftwaffe rapidly gaining air superiority. It is accepted that Stalin's purge of the Soviet high command go far to explain Russian disasters in the first days of the campaign. However, Nazi Germany failed to achieve her aims because of gross underestimation of Soviet manpower and resources. Before the commencement of the Soviet counter offensive three million Russian prisoners of war had been taken and twenty thousand tanks, twenty eight thousand guns and fifteen thousand aircraft had also been lost.

What was good enough to beat Poland and France was not good enough to conquer Russia whose resistance more than equalled the German challenge. France had been ideal country for armoured forces, but not Russia. The Soviet Union could not be defeated in a single campaign. The space was too vast, and the going too difficult. At the end of 1941, lacking hard intelligence about the Red army, their equipment, and deployment tactics, Hitler was left wondering what was holding the Red Army together. It was still strong and had nine million men of military age left, enough for four hundred divisions. Germany could not match those numbers, and a sustained battle of attrition strongly favoured Russia. German Intelligence reported that over six hundred tanks a month were coming out of the Russian factories, and when told of this, Hitler slammed the table, saying it was impossible. He refused to believe what he did not want to believe.

Perhaps Moscow could have been reached faster. But as a pre-condition to any move on Moscow, German flanks had to be secured and adequate transportation and supplies made available. This was not apparent at first, only at the end of September, and by then the Germans had run out of time and good weather. It is now accepted that by diverting the Panzers South to help Rundstedt take Kiev, the Germans lost valuable time. But, long before winter arrived the chances of victory had diminished.

The campaign in France had illustrated the decisive importance of maintaining speed. It is accepted that Russia is far larger than France, and the long distances took a terrible toll on German logistics, but this only reinforces the importance of reaching Moscow as soon as possible without any deviation. Guderian and others knew it, but Hitler ignored them. He did not understand military operations and strategy. Going for Moscow in August rather than October may have changed things dramatically, but would its capture have forced the Russians to surrender? The capture of Moscow would have hampered the Soviets' ability to move supplies, but the question of whether the capture of Moscow would have brought about the collapse of the USSR is questionable.

Proponents of the single thrust on Moscow tend to focus on best-case scenarios. They argue that the Germans were in a position to defeat the Soviets between July and October 1941, and had golden opportunities to inflict fatal damage on the disorganised Soviet forces. They would then have been in a strong position to negotiate peace with the Soviets after depriving them of their major resources. But, they stopped on their own initiative, and lost those opportunities. The capture of Moscow would have been only a symbolic victory for the Wehrmacht and a very costly one.

Once captured then what? Moscow would have been extremely difficult to garrison and require fresh troops, which were not available. Moscow would have been a hollow victory. The Red Army would still have been a force to contend with and the Germans would have been utterly exhausted. German planning completely failed to take into account what they would do if the USSR didn't collapse on cue: there were no plans to fight in winter, and the German economy wasn't even on a war footing able to provide the weapons and supplies that would be needed.

The relocated Soviet factories were turning out the materiel that made Soviet superiority decisive. The Germans had just one plan, and when it failed, their final defeat was inevitable. The proponents claim that instead of wasting time and resources in moving against Kiev, the fall of Moscow would have been decisive. It is difficult to speculate what Stalin would've done in this event. Perhaps he would've cut his losses and attempted to re-organise the Soviet Union, and perhaps grind Germany down in an unending war of attrition. But the question remains, "would Stalin have agreed to Hitler's demands even if Moscow had fallen?"

Hitler vainly stretched his forces too thin. The Africa Corps, the costly invasion of Crete and the Balkans campaign all detracted from the German effort of Barbarossa as well as the delay in its execution. There were German fronts along the coasts of Norway, Denmark, Holland, Belgium and France. German troops were also deployed in North Africa, The Balkans and Greece. Germany was quite incapable of sustaining this effort. Step by step, Hitler had led the German people and armed forces deeper into the realms of fantasy. Hitler's over-confidence stemmed from the crushing defeats that his army had previously administered. But, they could not achieve the impossible. Stalin is also not excused in this regard for despite being warned by both American and British intelligence, he simply dismissed all reports of German troops massing on the eastern border.

On the other hand, could this also support the Soviet first strike scenario?

So the two questions posed are:

*Was the Soviet Union about to invade Europe?

and

*If they were not, then why did Germany invade Russia?

The answer to question 1 is a very probable no.

Stalin had just completed the purges of his army, and had been largely humiliated at the hands of tiny Finland. He would need to be utterly delusional about the state of his army to engage on an attack on Germany at that time. And he was too crafty to contemplate such an adventure. Also, if the Red army was in a high state of battle-readiness, then why were they defeated so easily at the beginning of the war? What then is the answer to the second question - why did Germany invade Russia? The probable answer was not communism, but oil. Germany had no oil reserves and the Romanian oil fields were constantly under threat. That is why Germany was also fighting in North Africa and why in 1942 they went flat out for the Caucuses, not Moscow.

There are many factors and events that contributed to the failure of Operation Barbarossa. It is fair to say that not one particular issue contributed to the failure but a culmination of many events. The German army plunged into Russia under the impression that there were two hundred Russian divisions in total; only to soon discover that there were over four hundred divisions. The Germans also knew that the Russian roads were inferior for their vehicles and that the Russian railway tracks were a different gauge, yet logistics planning hardly took these factors into account before the invasion. They also failed to take into account Russian tenacity, the harsh weather and the terrain. Stories of this Russian tenacity spread widely among the Germans. Russian women had even taken up arms and troops would find teenage girls dead on the battlefield still clutching weapons.

Hitler truly was confident that the delay in launching the invasion was of no consequence, but he had no way of knowing just how fiercely the Russians would oppose him. So, the combination of all these factors led to the failure. Undoubtedly, poor preparations by a military force that prided itself on thoroughness was a major factor that contributed to failure. For example, Germany's arms production in the second year of war had barely increased, whereas in the United States, Britain, and the USSR it had almost doubled over the same period. The consequences were that the Wehrmacht invaded Russia with fewer artillery pieces than had been available against France. Many infantry and motorised divisions had to be equipped with foreign vehicles. The Germans are inclined to blame their defeat on "General Winter". But this is a poor excuse. Germany had control over practically all European industries, including textile industries that were capable of supplying the Wehrmacht with anything necessary.

The invasion of Russia was doomed to fail from the beginning due to the fact that the Germans were unprepared and extremely overconfident. When the invasion was launched, Hitler was convinced that he would achieve his objectives within eight weeks. That was an arrogant blunder. Perhaps Hitler should not be held as a scapegoat. However, he was very much responsible for the failure of Operation Barbarossa, and consequently the war. He had an inflated opinion of himself and underestimated the enemy. He thought Germany could take on the world, and he made stupid errors to satisfy his pride. But what if he had triumphed and he was within an inch of doing so. Had Hitler triumphed, Germany would have possessed Russia's limitless natural resources. Britain would very probably have yielded, and Washington may have acknowledged Berlin's dominance of Europe.

The Americans may then have been forced to fight in a war between continents in which Hitler had far greater resources of manpower and industrial might than did the world's last great democratic power.

The campaign was, indisputably, therefore, the most critical of the war. What might have been is pure hypothesis. What did happen was the total destruction of German military power. In the end it had all been in vain!

Hitler's Napoleonic dream had become the impossible dream; operation Barbarossa had failed!

~EIGHT~
PEARL HARBOUR

The devastating blow delivered by Imperial Japan at Pearl Harbour on the 7th December 1941 effectively introduced the Americans to "total war." Nothing experienced by any nation in both world wars even remotely compared with the catastrophic naval defeat suffered by the United States on that fateful day in world history. Their unsuspecting fleet was caught completely unawares when over three hundred and fifty aircraft from six Japanese carriers decimated the US Pacific Battle Fleet, inflicting the worst military defeat suffered in America's two hundred year history, leaving them totally bewildered and completely astonished.

In terms of military achievement and for sheer audacity the raid on Pearl Harbour is without parallel and merits a special place in naval warfare. In a masterful stroke, a Japanese carrier task force steamed undetected across five thousand kilometers of open sea to launch the greatest combined air/sea operation of all time. Japan's success at Pearl Harbour was made possible due to superb training and remarkable duplicity. As a result she was placed in a position of overwhelming strength throughout the Far East.

In a wave of conquest unmatched in both world wars, several operations were launched simultaneously against British, Dutch and American possessions stretching ten thousand kilometres. One by one the great bastions of white supremacy and white prestige fell before the Japanese onrush. One of the few international statesmen with the vision to realise the consequences of this development was none other than South African Prime Minister Jan Smuts.

Even as the attack on the U.S. Pacific Fleet was taking place at Hawaii, Japanese troops were already landing on the Malay Peninsula and preparing to advance overland to Singapore, later to capitulate after Japan inflicted on the British army the most humiliating defeat in its history. This was accomplished by skillfully executed tactics that almost entirely eliminated their British adversaries as combatants in the Far East.

Pearl Harbour proved to be a fatal blow to western imperialism and facilitated the coup de grace that was administered at Singapore. On the 10th of December, just three days after Pearl Harbour; HMS Prince of Wales and HMS Repulse under the command of an admiral totally disbelieving in air power, also made history by becoming the first capital ships to be sunk from the air in open sea. In an era when the battleship was considered, with unshakable conviction, to be the ultimate maritime weapon, the fate of these two great ships proved once and for all that powerful battleships were not immune to destruction from the air. With great reluctance it was now accepted that naval air power had come of age. Ironically, this was only accepted after every allied battleship in the Far East, ten in all, were either sunk or put out of action.

The story of Pearl Harbour has for too long been over-simplified. The question of why Japan caught the United States napping is exceedingly controversial and one must constantly keep in mind that nothing takes place in a vacuum. The background to this calamitous event goes back to the end of World War One when the US and Japan became increasing aware that war between them was inevitable and both countries planned for such an event. Japan believed the Western Powers were hostile and looked down on them as being an inferior race. Relations became strained due to US opposition to Japan's imperialism and expansionist policies; thus reducing Japan's dependency on the US for supplies, in order to create a *"Greater East Asian Co-Prosperity Sphere"*.

This political concept was to promote a self-sufficient bloc of Asian nations led by Japan, free of Western colonialism and influence. In reality it was nothing more than a scheme to control countries in which puppet governments would manipulate its economies and peoples for the benefit of Imperial Japan. Without these economic resources and raw materials Japan would never become a dominant power in the region.

Japan's invasion of China in 1937 and the Tripartite Alliance formed in 1940 between Japan and the other two Axis dictatorships, namely Nazi Germany and Fascist Italy, resulted in America applying economic sanctions that included an oil embargo. Tensions continued to escalate when President Roosevelt moved the US Pacific Fleet from California to Hawaii in the hope of discouraging Japanese aggression in Asia. The Commander in Chief Admiral Richardson told Roosevelt this move was not a deterrent but an incitement to Japanese aggression. Roosevelt sacked him.

In July 1941 tensions further intensified when Japan occupied French Indochina, causing the US to impose a total trade ban and freeze Japanese assets. Facing economic collapse and withdrawal from recently acquired territories, Japan could either agree to US demands and restore normal trade practices or use force to gain access to resources. By November 1941, the Japanese though still continuing to negotiate with the United States, had come to the conclusion that further talks towards the lifting of sanctions were futile. Their only alternative was war, or the abandonment of their objectives to dominate the Far East.

On this basis, if there had to be war, then the present time seemed opportunistic for Japan. At this stage the German armies were threatening both Moscow and Egypt. A great part of the United States Navy was engaged in an undeclared war in the Atlantic against the U-Boat menace. Therefore, the Japanese military leaders prepared to exploit this deepening world crisis and seize the mineral rich territories in Asia; which, they considered to be theirs by right and were prepared to challenge the Western powers for possession.

Other factors prevalent in November 1941 were that Great Britain was in such a defensive position that, even if she survived, her entire war-making potential would be exhausted in the desperate defence of her homeland never mind her empire. Potential enemy forces that could be deployed in the Far East, particularly in the air, were insufficient to prevent the fully trained and mobilised forces of Japan from achieving her goals. Hitler's Germany had neutralised any threat from the Soviet Union, who appeared to be on the verge of collapse. In accordance with this scenario, the Japanese Army was given the primary responsibility for conquering Malaya, the Dutch East Indies, and Burma. Simultaneous attacks would be launched on Java and Sumatra with the intention of isolating Australia and New Zealand. The navy would play a supporting role by embarking on several combined operations. But in doing so, Japanese naval forces would most probably have to contend with the American Pacific fleet anchored at Pearl Harbour.

Admiral Yamamoto the commander of Japan's Combined Fleet proposed a sudden and paralysing knockout on the U.S. fleet at Hawaii. Yamamoto believed if there was to be war with the U S, Japan would have no hope of winning unless the US Fleet in Hawaii was destroyed. He calculated that the United States would be so weakened by this proposed attack that she would be unable to mobilise sufficient strength to go over to the offensive for about two years. By that time, the conquered territory would be fortified and Japanese stubborn resistance would undermine American determination to continue the war. The Japanese speculated that the United States, in the face of potentially unacceptable losses would probably compromise, thus allowing Japan to retain a substantial portion of her territorial gains. When one considers what happened years later in Vietnam, perhaps this hopeful speculation was not entirely unfounded.

However, returning to 1941, Yamamoto had been strongly influenced by the successful British Fleet Air Arm operation at Taranto the previous year when three Italian battleships had been put out of action by only twenty-one obsolete Swordfish torpedo planes launched from a single aircraft carrier. Consequently, Yamamoto planned and organised a massive Strike Force consisting of six aircraft carriers, two battleships, three cruisers and eleven destroyers to attack Pearl Harbour in a pre-emptive strike. The Philippines would be seized to deny the US a base for possible retaliation.

On November 26th, the strike force left Japan and proceeded in utmost secrecy on a mission intended to deal a fatal blow to the U.S. Pacific fleet. Just prior to launching their aircraft at dawn on December 7th from a point three hundred and fifty kilometers north of Hawaii, the Japanese received an intelligence report that no American aircraft carriers were at Pearl Harbour. This was a disappointment, but, there were still eight battleships in the harbour anchored in pairs at Battleship Row and these were the primary targets. As previously noted, the battleship at this time, was still considered to be the principal capital ship and the aircraft carrier, merely a support vessel.

Diplomatic negotiations with the US continued and if successful the ships were to return to base and no one would have been any the wiser. Reports confirmed that no defensive measures were in place indicating complete surprise. An hour before the Japanese air strike on Pearl Harbour, the first action of the Pacific war took place when a US Destroyer sighted, attacked and sunk a Japanese midget submarine outside the harbour entrance. Paradoxically, the first shots fired to launch the Pacific war, came from the United States and not from Japan, and the first casualties were Japanese not American. The Destroyer captain promptly reported this encounter with the Japanese submarine, but unfortunately for the Americans, no one took any interest. If the alarm had been raised it would most certainly have provided sufficient time for Pearl Harbour fighter aircraft to take to the air and enabled American ships to raise steam and make for the open sea.

Ninety six warships, of all shapes and sizes were in harbour Anti-Aircraft guns were manned but ammunition was locked up and the keys held by officers, some of whom were ashore. Army OC Lt Gen Walter C. Short had received a signal from Washington stating "*hostile Japanese action at any moment*".

Admiral Kimmel, Commander in Chief, Pacific Fleet, was notified that the Japanese were burning their codes and had changed call signs twice during the month. Of major importance he was also informed that Naval Intelligence was unaware of the location of the Japanese Carriers

Amazingly, Admiral Kimmel had received a dispatch from his military chiefs in Washington on 27th November stating clearly and comprehensively - *"consider this to be a war warning"*. This dispatch, delivered ten days before the Japanese onslaught, informed him that diplomatic efforts had failed and Japan was expected to make an aggressive move within a few days. General Short, was of a similar frame of mind as Kimmel; he did not even remotely contemplate that disaster was about to strike and openly boasted that Pearl Harbour was *"the best defended naval base in the world"*.

He had become obsessed and totally preoccupied with preventing acts of sabotage. Therefore, he ordered the aircraft to be close-packed on airfields rather than dispersed and ready for action. Thus, on the morning of December 7th the aircraft for the defence of the Pacific Fleet at Pearl Harbour were huddled together and no ammunition was readily available. American air strength at Hawaii was about to be completely shattered, and two thirds of the entire American aircraft in the Pacific theatre destroyed.

The military commanders in Hawaii could not even conceive the type of attack that the Japanese would launch. Not at any time did they envisage an air strike aimed at the destruction of the U.S. Pacific Battle Fleet. So, when the attack did come, it was a traumatic shock to them. General Short actually thought this was an air raid intended to soften up the defences prior to the Japanese invasion of Hawaii. Meanwhile the military authorities in Washington expected their men on the spot to take full defensive measures at Pearl Harbour. They anticipated that submarines would be sent out on patrol and the battleships with their protective vessels would be at sea deployed in readiness for war. Anti-aircraft defences were expected to be placed on full alert around the clock.

This was not an unreasonable deduction; after all, it is the responsibility of the military commanders on the spot to prepare for the worst possible contingency whether in peace or war. However, the commanders at Hawaii convinced themselves that the Philippines or Malaya would be the prime Japanese targets, not Pearl Harbour. They, therefore, took no action in preparation for war. Their view was that Japan would never deliberately initiate war with the United States. The idea was considered too preposterous. For the defenders of Pearl Harbour, events were about to go irretrievably and horribly wrong.

To this day, it is quite inconceivable that Pearl Harbour was not put on the state of high alert that it should have been. The senior military commanders concluded, despite the many warnings, that there was no reason to believe an attack was imminent. Therefore, anti-aircraft guns were unmanned and most ammunition boxes, in accordance with peacetime regulations were kept locked.

December 7th was a Sunday morning and many officers and crewmen were allowed to go ashore despite the war warnings. Taking all of this into consideration it is understandable why the Japanese surprise attack was so complete and how they were to rapidly gain air superiority during the crucial first phase of the strike.

The Army Radio Station detected a large number of aircraft 220 kilometers away to the north heading for Oahu. The operators tried to raise the alarm but as the carriers were at sea and a flight of B-17s were expected from the US mainland nobody at the Info Centre was interested.

When the attack did start, Japanese Zero fighter planes made low level strafing sweeps, shooting up aircraft as well as ground installations. Thus permitting coordinated attacks on the battleships moored in battleship row by dive bombers, level bombers and torpedo planes. The primary attack led by Commander Fuchida consisted of one hundred and eighty four aircraft and caused the greatest amount of damage in a very short space of time. From endlessly repeated practice runs and the meticulous study of maps and models of Pearl Harbour, every Japanese pilot knew exactly what they had to do and aimed at their targets with cool precision.

Struck by three torpedoes, *Oklahoma* capsized almost immediately; trapping below decks four hundred and fifteen men. One aerial bomb blew up the forward magazine of Arizona, while another dropped down the funnel and exploded in the engine room. The ship went down with over one thousand men out of a crew of one thousand five hundred. A torpedo and two bombs hit the Nevada. The Japanese then launched their second carrier strike to complete the work of the first wave.

Numerous aircraft were observed over Ford Island as smoke and debris shot into the air. Japanese aircraft skimmed low over the harbour accompanied by muffled explosions. Within five minutes, four battleships had been holed or severely damaged below the waterline by torpedoes. Dive bombers smashed the decks, bridges and gun turrets to finish off the job. Within half an hour Battleship Row had been devastated and out of one hundred and forty eight first line aircraft one hundred and twelve were destroyed. The first attack wave withdrew having lost only nine aircraft

One hundred and seventy one aircraft attacked in a second wave consisting of thirty six fighters and one hundred and thirty five bombers to attack smaller warships, airfields and targets of opportunity. This was far less effective than the first attack and more aircraft were lost due to the Americans being better prepared.

A third wave attack which may have devastated what remained of Pearl Harbour was called off by Admiral Nagumo, despite being urged by his staff to go ahead with it. Fuel storage tanks which held more fuel than Japan had in its strategic reserve were still intact. As were dockyard facilities and command control structures. Admiral Nagumo, however, decided to withdraw because he believed the objective, the neutralising of the US Pacific Fleet, had been achieved.

Further considerations relating to another attack were the probable disproportional increase in aircraft losses, and the unknown whereabouts of the US Carriers. While this was taking place, a Japanese midget submarine succeeded in penetrating the harbour, passing through the entrance gate that had been carelessly left open. This submarine was sighted firing a torpedo at the Oklahoma and was promptly depth charged and sunk.

When the smoke lifted over Pearl Harbour, more than two thousand four hundred Americans were dead and one thousand three hundred more were wounded. Some two hundred and thirty aircraft had been destroyed or heavily damaged. The Japanese lost just twenty nine planes and less than one hundred men. Eighteen ships had been hit and Battleship Row was a shambles. The Arizona and the Oklahoma were a total loss. The California and the West Virginia were sunk at their moorings. The Nevada, the Maryland, the Tennessee, and the Pennsylvania were all heavily damaged.

Once the Japanese aircraft withdrew, General Short prepared for the invasion that he thought was bound to follow. Now the alertness and preparedness that could have provided such a hot reception for the Japanese was brought to life after the enemy departed. Nervous sentries opened fire at anything that moved and friendly planes were shot down in the belief this was a renewed aerial assault.

On 8th December President Roosevelt at a Joint Session of Congress delivered his famous "Day of Infamy speech". In less than an hour the US was officially at war with Japan; declaring that a state of war existed between the United States and the Empire of Japan. The President then proceeded to waste little time in sacking Admiral Kimmel and General Short for a gross neglect of duty. The entire blame was laid squarely on their shoulders for ignoring all the warnings and indications of obvious Japanese intentions. Pearl Harbour resulted from a vast combination of interrelated factors. On the one hand, bountiful human errors, false assumptions, and a vast store of intelligence information badly handled. On the other, precise planning, tireless training, fanatical dedication, iron determination and tactical excellence. It was a daring and brilliant naval operation.

The military supremo's on both sides were far off the mark in their evaluation of the military threats posed by each other. Japan gambled on the US agreeing to negotiate when faced with a sudden and massive defeat. The effect was the exact opposite. It united US public opinion to declare war on Japan. A major flaw in Japanese naval strategy was a belief that the ultimate Pacific battle would be fought between battleships in Japanese waters. Failure to attack shore facilities enabled the Pacific Fleet to continue operating from Pearl Harbour. Thus the threat to Japan's Eastern flank still remained. Japan didn't get the twelve month respite she needed to secure her gains and six months later the tide of war turned against her at the Battle of Midway

The presence of the US Pacific Fleet in Hawaii failed to deter Japanese aggression and against all expectations resulted in a great tactical victory for Japan. The US believed the Imperial Japanese Navy was too widely committed and thus unable to attack the Pacific Fleet. This air of invincibility was well and truly shaken to the bone.

Loss of her battleships forced the US to use carriers and submarines to take on the Japanese at sea which was to reverse the Japanese advance and lead to her defeat. The battleship, regarded as the most powerful and significant element of naval power was relegated to a support role. One Japanese Naval Commander summed up the result when he remarked *"We won a great tactical victory at Pearl Harbour and thereby lost the war"*. Embracing so much so suddenly so unexpectedly and so spectacularly; it still seems inexplicable and mysterious.

What has been addressed so far may be considered to be the traditional view of which most historians are familiar. But, in probing a little deeper and analysing other aspects of this world-shattering event, several theories emerge. Considering the explosive force that Japan deployed in her conquest of South-East Asia, one is stunned by the irresponsible propaganda that was put out in the American press about Japanese military strength.

The West hopelessly underestimated Japan's military capacity and offensive capability. In their ignorance they claimed as a well-established fact that due to a deficiency of vitamin C the Japanese lacked good vision; hence, they would make poor air pilots and their naval personnel would be no match for their Western adversaries, particularly after sunset. What is more, the Americans held the Japanese in utter contempt, viewing them as funny little creatures with buckteeth and horn-rimmed glasses covering slanted eyes. They were viewed as slow brained, inefficient, and incapable. The Americans assured themselves that Japan was virtually bankrupt and exhausted from being bogged down in China. They considered Japan to be one hundred years behind the times, and if she engaged in a major conflict, her fragile economy would simply shatter. This is quite amazing, considering that on the eve of Pearl Harbour, more than half of Japan's budgetary expenditure for the year 1941 went on armaments.

The British and American press maintained that the Japanese navy consisted of only four small aircraft carriers and two hundred aircraft that could not possibly meet the requirements of a modern war. Perhaps, this only goes to prove how successful the Japanese were in concealing their military strength from foreign observers. Nevertheless, these ludicrous assumptions accurately reflect the Western underestimation of Japanese air power and this is best illustrated by the reliance they placed upon the antiquated Brewster Buffalo fighter plane. The Americans actually thought this aircraft was far superior to anything that the Japanese could put in the air.

It was propagated quite arrogantly, that Japanese planes and Japanese airmen were no match for their British and American counterparts and that Japanese aircraft were poor imitations of outdated allied aircraft. In actual fact, as the Western powers were soon to realise at a terrible cost, the Japanese aircrews were superbly trained for a whole range of offensive and defensive missions. The daring and expertise of first rate experienced airmen contributed immeasurably to Japan's early conquests.

In 1941, the Japanese had the best naval arm in the world and were the pioneers of the large carrier striking force, and by the time of Pearl Harbour. Japanese pilots had undoubtedly attained the world's highest bombing standards and were extremely skilful. They had torpedoes that were of a far superior design and much more effective than the Allied torpedoes.

The Allies mistakenly thought the Japanese were incapable of producing anything like the Spitfire or the Messerschmitt. They soon received a monumental shock. The Japanese "Zero" fighter was scornfully described by an American aviation expert as "nothing more than a light sports plane". While in fact, in the hands of highly trained Japanese carrier pilots, the Zero was a most effective and deadly weapon.

The Zero was faster than any opposing plane of the period, and it could out manoeuvre, out-climb, out-range and packed a heavier punch than opposing aircraft. At the start of the Pacific war, the Zeros quickly gained air control over any battle area. American and British aircraft fell like flies before the agile Zero fighters; especially the Brewster's that literally flew on suicide missions against the Zeros.

Not everything however, went entirely the way of Japan. From a realistic perspective, they also made mistakes. The most notable being the refusal to order a third strike at Pearl Harbour. They could easily have destroyed the repair facilities and, more importantly, the oil storage tanks. This vital target stored four and a half million barrels of oil and escaped undamaged. The Japanese carrier commander later explained that he thought the Americans still had a large number of land based aircraft in operational condition at Hawaii and considered it too risky to remain within their range.

Perhaps this was being over cautious, or was it being pragmatic? After all, a successful counter strike by the Americans against the Japanese carriers would have reversed the entire situation. The defenders at Pearl Harbour, although caught napping, had quickly recovered as had been observed by the much heavier losses incurred during the second strike. Undoubtedly, a third strike would have met with strong opposition and would possibly render Japanese losses disproportionate to any additional damage inflicted.

However, in effect, this was a grave error because the loss of the oil supplies at Pearl Harbour would have hindered American naval operations in the Pacific far more than the damage done to the fleet; considering that a single destroyer steaming at full power uses up its fuel supply in thirty to forty hours. This gives some idea of the staggering needs of an entire fleet. Also it must be borne in mind that Hawaii produced no oil and every drop had to be transported almost four thousand kilometers from the mainland. This accumulated fuel reserve would have taken many months to replace. Without it, the Pacific fleet could not have been able to operate from Pearl Harbour and the surviving warships would have been entirely immobilised and incapacitated. The failure to destroy the fuel tanks and service facilities reflected Japanese preoccupation with tactical rather than logistical targets.

Perhaps it was the very ease of their Pearl Harbour operation that threw the Japanese off-balance. Against all expectations and without as much as even a scratch on any of their ships they possibly lost sight of their primary objective, which was to inflict the maximum amount of damage on Pearl Harbour. They now thought only of the minimum damage to themselves. Furthermore, and this was to prove critical to the outcome of the war, escaping damage from the attack, were the three all-important American aircraft carriers in the Pacific, the Lexington, the Enterprise and the Saratoga. In failing to seek out and dispatch the American carriers, Japan committed its first and probably its greatest strategical error of the entire Pacific conflict.

Allegations have been made that some members of Roosevelt's administration, including the President himself, had advanced knowledge of the attack but purposely ignored them and deliberately exposed Pearl Harbour to force the US into war. The number of intelligence reports from diverse sources that Pearl Harbour was to be attacked, creates suspicions that this may have been the case. The alternative is dereliction of duty and gross incompetence at a very high level. Did the Americans actually know of the forthcoming attack on Pearl Harbour? Bearing in mind that Washington had managed to "break" the Japanese diplomatic codes. But before proceeding with this line of thought, the sequence of events leading up to Pearl Harbour and the immediate aftermath needs to be recapped.

Saturday, December 6th - U.S. intelligence intercepts a Japanese coded message instructing their embassy staff to destroy all codebooks and prepare to depart. This information was passed on to President Roosevelt, who commented *"This means war"*. So, there is little doubt that a Japanese attack was imminent.

Sunday, December 7th - The U.S. War Department sends out an alert but uses a commercial telegraph because radio contact with Hawaii is faulty. The telegraph is not marked urgent and is only delivered four hours after the attack had begun. Pearl Harbour radar operators detect Japanese aircraft approaching, but the report is disregarded on the assumption that the incoming aircraft are American. At 07:53 hours still on December 7th, the first Japanese assault wave commences the attack, taking the Americans completely by surprise.

Monday, December 8th - The United States and Britain declare war on Japan announcing December 7, *"a date which will live in infamy"*.

Thursday, December 11th - Germany and Italy declare war on the United States. The conflict has become global with the Axis powers;

Japan, Germany and Italy, ranged against the Allied powers, America, Britain, and Russia.

Pearl Harbour revisionists, in pursuing the "conspiracy theory" dispute the official judgment that the unreadiness of US forces when the Japanese attacked Pearl Harbour was solely the fault of the local commanders. They claim that Admiral Kimmel and General Short were made scapegoats. The conspiracy theory holds that President Roosevelt specifically planned to bring about an attack on Pearl Harbour and intentionally withheld information on the pending attack from the Hawaiian commanders. He set them up by denying them intelligence reports and misled them into thinking that negotiations with Japan were progressing favourably. The revisionists hold President Roosevelt personally responsible for the debacle at Pearl Harbour and accuse him of using the incident as a means of getting the United States into the war.

They maintain that he deliberately provoked the attack when he cut off Japan's supply of oil, which was intended to force them into war with the US. They state that the Pacific Fleet was intended to play the part of the goat tethered as bait to the Japanese tiger. The question posed by them is; "why did Roosevelt order the Pacific fleet to be transferred from the West Coast to its exposed position in Hawaii and why did he overrule objections by naval commanders that there was inadequate protection for the fleet at Pearl Harbour"?

In relocating the Pacific Fleet to Pearl Harbour, as opposed to San Diego, Roosevelt effectively sealed the fate of the battleships, because the Japanese considered this move as tantamount to a dagger pointed at their throats. According to the "conspiracy theorists", the U.S. Navy had positive proof that the Japanese were planning to attack Pearl Harbour before Christmas 1941. The governments of Britain, Holland, Australia, Peru, Korea and the Soviet Union were aware that an attack on Pearl Harbour was coming and had warned the US. Therefore, Roosevelt and his inner circle knew of the attack, allowed it to happen, and covered it up.

This postulates a further point. If the Americans did know about the attack then why did they not intercept and smash the Japanese carrier strike force off Hawaii? This would still have provided the American

President with a reason to declare war on Japan and would have saved the fleet at Pearl Harbour. The revisionist explanation to this point is that Pearl Harbour was not about war with Japan, Roosevelt's objective was war with Germany, and he needed the Pearl Harbour attack to provoke Hitler into declaring war on the US. This could only be achieved by baiting Hitler into thinking the U.S. could be beaten.

If Japan's fleet were destroyed by a pre-emptive U.S. strike, this would defeat Roosevelt's purpose; inasmuch that it would have been obvious suicide for Hitler to declare war on America if his major ally, that is, Japan were crippled. Direct provocation had failed when the President in his capacity as Commander in Chief of American armed forces had ordered the US navy to attack German U Boats in the Atlantic. Hitler refused to rise to the bait. Roosevelt anticipated that if Japan attacked and scored a decisive victory over the U.S., Germany would declare war on America in terms of the Tripartite Pact.

So, according to the "conspiracy theory", the objective of the President and his senior staff was to manoeuvre Japan into firing the first shot in order to leave no doubt in anyone's mind as to who are the aggressors. The American President and his chief advisors had construed a fatal trap to entice Imperial Japan. They claim that this is supported by the behaviour of General Marshall, the head of the US. Army who was the only officer empowered to issue a full alert warning. However, on the morning of the attack he had gone out horse riding. His intelligence staff that were by this stage becoming quite frantic in the knowledge of the impending attack, were informed that he could not be contacted and had left orders not to be disturbed.

It may be appropriate to recall that Hitler at D-Day had also left instructions not be awakened which resulted in the failure to release his panzers at a critical time, and possibly cost him the war. Be that as it may, when Marshall was eventually tracked down by his intelligence officers, he stubbornly refused to accept their reports and they had difficulty in persuading him that the Japanese were about to strike. The revisionists state that after hearing of the attack at Pearl Harbour, and before any damage reports were received, the

American President called Lord Halifax at the British Embassy and told him: *"Most of the fleet was at sea, none of our newer ships were in harbour, and they only destroyed a lot of old hardware and did us a favour."*

The proponents of the conspiracy theory claim that FDR sold out his country in order to be President for life. They consider him to be a traitor for manoeuvering Japan into war with the U S; a traitor for sacrificing American lives, and a traitor for usurping the Constitutional power of Congress. They claim that four days before the attack, FDR could have sent telegrams of condolence to the families of the servicemen he was allowing to be killed. The Day of infamy actually belongs to him.

That is the Conspiracy Theory! The actual events are that from coast to coast, the Americans initially reeled in traumatic shock after Pearl Harbour. But this subsequently turned to fury. The Japanese attack united one hundred and thirty million Americans in a relentless determination to push the war against all odds to final victory. On the bridge of the U.S. aircraft carrier Enterprise as she headed back to Pearl Harbor on December 9th, Rear-Admiral Halsey, at the sight of the wrecks at Ford Island uttered: *"Before we're through with them, the Japanese language will be spoken only in hell"!*

When the news of the *"sneak attack"* was broadcast to the American public, it sent a shockwave across the nation; resulting in a tremendous influx of young volunteers into the U.S. armed forces. The attack also united the nation in a cyclone of wrath and effectively ended any isolationist sentiment in the country. December 7th mobilised all American resources and raised a mighty wave of indignation across the United States, which would eventually break on Japan with colossal force. The American people were aroused, as no other event in their history had ever done and a lot of people went absolutely berserk. They went as far as chopping down Japanese cherry trees and even slaughtered Japanese deer in New York Central Park. To every American the Japanese were the embodiment of the evil enemy, the cowardly stab-in-back aggressors. The wave of American hatred against Japan was staggering.

What did Japan actually achieve by the attack on Pearl Harbour? From a military viewpoint, Yamamoto actually removed an albatross from around the neck of American naval policy; because many American admirals had never accepted the aircraft carrier as a capital ship. They believed that the coming war would he fought in much the same manner as the last, where fleet battles would again consist of gunnery duels between lines of heavy ships. To them, big guns, not aircraft were still the decisive naval weapon.

Pearl Harbour effectively cancelled American line doctrine and facilitated a shake-up of the navy high command. The fleet carriers, of necessity became the major capital ships and forced the Americans to adopt carrier warfare, which in the long run was to give her victory. The battleship was now relegated to secondary duties such as offshore bombardment and convoy protection. The carrier now took over as the major strike weapon. Big gun battleships could now only operate after command of the air was assured. Therefore, Pearl Harbour was to prove much less of a disaster than it first appeared. The old US battleships that were sunk were in any case too slow to stand up to Japan's newer and faster battleships and too slow for the American fleet carriers. The Americans could also take consolation, despite severe losses, that Pearl Harbour's repair facilities were practically untouched and would restore most of the damaged ships to service.

All but the Arizona and Oklahoma were raised, repaired, and renovated, and these ships would later engage in amphibious operations. Meanwhile, their temporary loss proved to be a bonus for the Americans, as they freed thousands of trained seamen for use in the carriers.

The eventual fate of the major ships involved at Pearl Harbour was as follows:

Pennsylvania had all her repairs completed in a few weeks at Pearl Harbour and was back in action by the 2nd January 1942. She went on to win battle honours and survived the war.

Maryland also had all her repairs completed at Pearl Harbour and was soon back in action. She also went on to win battle honours and survived the war.

Tennessee returned to the United States for extensive repairs and rejoined the fleet in February 1942. Besides winning battle honours she also received a navy unit commendation and survived the war.

Nevada also returned to the United States for extensive repairs and rejoined the fleet in April 1942. She then underwent a thorough overhaul plus modernisation programme and went on to see action in the Normandy landings, Iwo Jima, Okinawa and Japan itself, winning many battle honours and surviving the war.

California also underwent a similar modernisation programme as the Nevada and rejoined the fleet. Besides several battle stars she sank a Japanese battleship later in the Pacific. She also survived the war.

West Virginia was not dissimilar to the *Nevada* and *California* and was present at the surrender of Japan in Tokyo Bay on September 1945.

Oklahoma - great efforts were made to salvage her, but she was finally decommissioned in 1944.

With the exception of the Arizona and the Oklahoma, all the US. Battleships at Pearl Harbour were eventually repaired and returned to service.

The fate of the Japanese ships is somewhat different. Four of the aircraft carriers were sunk six months later at the Battle of Midway. The remaining two carriers and battleships were sunk in 1944 at the Battle of Leyte Gulf. By an ironic twist of fate, nearly all the victims survived to fight again, and all the attackers ended their days at the bottom of the Pacific Ocean; the very same ocean that they sought to dominate.

It now becomes obvious that the Japanese miscalculated and did not think through the strategic problems associated in a war with the United States. Even Yamamoto, the architect of Pearl Harbour, knew that Japan could not win a protracted war and had hoped for a compromise peace with his country's great Pacific rival. But, Yamamoto's Pearl Harbour operation only served to provoke the people of the United States to such fury that nothing short of unconditional surrender would satisfy them. There would be no negotiated peace and sixteen months after Pearl Harbour the Americans had their revenge. Retribution was swift; Yamamoto was ambushed by U S aircraft in the south Pacific and killed

Returning to the "conspiracy theory", that infers President Roosevelt desired and deliberately engineered the attack to bring the United States into the war by the "back door'. After exhaustive research, neither evidence nor common sense justifies this viewpoint. The economic embargo was not a malicious attempt to bait Japan into war, it was intended to make Japan stop, look and listen. But, Japan was so determined to gain the raw materials of Southeast Asia that she was prepared to fight any power or even a combination of powers to secure them.

To give credence to the "conspiracy theory" one would have to accept that the Japanese were puppets for Roosevelt to manipulate at will, and this was most certainly not the case. The Japanese planned and executed the attack entirely on their own with no urging or assistance from President Roosevelt. The individual responsible for the attack on Pearl Harbour was Admiral Yamamoto not President Roosevelt. If there had been no Yamamoto, there would have been no Pearl Harbour.

In retrospect, it could be deduced that war between the United States and Japan was inevitable, Pearl Harbour or no Pearl Harbour. Japan was on the march towards what she earnestly believed to be her destiny. The Japanese propagated that they were merely adopting the same policy as the Americans had in the previous century with their *'Monroe Doctrine"* (America for the Americans). The Japanese were now saying Asia for the Asians; albeit under Japanese control. If the United States would stand aside, and allow the Japanese to grab the Asian territories, well and good; if not, the United States would have to accept the consequences. It was a catch 22 situation, and the price of peace was to give the Japanese everything they demanded.

In summation the disaster at Pearl Harbour was not due to any conspiracy; it was due to military incompetence. Human shortcomings, not betrayal allowed the Americans to be taken by surprise. The commanders in Hawaii failed to discharge their basic responsibilities on December 7th by not engaging the Japanese with every weapon at their disposal. They were clearly warned of war with Japan and were informed that the entire Pacific region was a danger zone. This evaluation is based on the incredible decisions made by the men on the spot. Admiral Kimmel refused to even consider the possibility of a carrier borne air attack, assuming this to be out of the question. No preparation was made to repel such an attack and Kimmel made an inexplicable error of failing to institute long-range aerial reconnaissance.

It later emerged that he had expected the army to perform this task although he knew it was the navy's responsibility. These planes would most probably have located the enemy battle squadron and enabled the fleet at Pearl Harbour to be alerted to the presence of the Japanese carrier force. The American fighter planes would have been airborne, the ground defences manned and the Pearl Harbour story would have been vastly different. In view of the gravity of the situation and the frequent war warnings that had been issued, this was indeed poor judgment. Kimmel did not even cover the northwestern approach to Pearl Harbour, even though frequent war exercises revealed this as the most likely area for an enemy carrier based air attack.

The disaster at Pearl Harbour occurred because its defenders were not alert and they lacked vigilance. They failed to detect an approaching enemy task force within three hundred and fifty kilometers off Hawaii. They failed to challenge Japanese patrol aircraft over Pearl Harbour. They failed to act on the reported presence of an enemy submarine, and they failed to respond to radar sightings of Japanese aircraft. Imagine what would have happened if British radar operators had been so complacent at the Battle of Britain. If they had, perhaps the war would have been over a year sooner and there would have been no need for Pearl Harbour.

Kimmel could only envisage a fleet action similar to the battle of Jutland, and advocated leading his substantially inferior battleship fleet deep into hostile waters to give battle to the enemy. It did not seem to occur to him that by the time he would have reached the combat area, his crewmen, who had never fired a shot in anger and who trained only in the most pleasant of waters, would have been exhausted. Japanese submarines would have whittled the American fleet down to size and he would have to give battle to the Japanese in their own hunting ground and on their terms when he was at his weakest. The Americans would not have been able to recover their ships as at Pearl Harbour and their losses would have been possibly twenty thousand men not three thousand.

Kimmel's policy planning had little basis of reality. No adequate plan of defence was made and months before the attack on Pearl Harbour, he allowed the Hawaiian National Park Service to block the installation of radar stations at vantage points. Perhaps Admiral Kimmel agreed that scenic views were much more important than an efficient radar station at Hawaii. A strange inconsistency arises out of Pearl Harbour. Kimmel and Short were dismissed. But, General Macarthur who also got clobbered in the Philippines, remained free from investigation. Macarthur had the advantage of being fully aware that Pearl Harbour had been attacked and knew that the enemy was coming in his direction. Yet Japanese aircraft also found his planes parked helplessly on the ground and obliterated his air force. The lack of air cover was the main reason for the fall of the Philippines.

Pearl Harbour will be debated as long as naval history is written. Far from being a strategic necessity as the Japanese were later to claim, the surprise attack on the Pacific fleet was in effect a short-term tactical masterpiece but a colossal long-term strategic blunder, and from a political viewpoint an absolute disaster. Never in military history has an operation proved more fatal to the aggressor.

What would have happened if the Japanese had launched a third strike, or had invaded Hawaii and perhaps had even caught the American carriers at Pearl Harbour?

This poses one of the most awesome *"IFS"* of the twentieth century because the entire story of World War Two would have changed. If the Japanese had used Hawaii as a springboard to engage the Americans on their long, virtually indefensible Pacific coastline, the consequences would indeed have been far-reaching.

Finding the Japanese fleet on their west coast and at the gateway of the Panama Canal, it is extremely doubtful if the Americans would have shown much interest in the war in Europe. The entire American war effort would have been geared against the Japanese forces on their doorstep. American public opinion would not have allowed a single GI to sail to Europe. Projecting this scenario forward, there would have been no Normandy landings because Britain could not undertake this alone.

If Germany was to be beaten then it would have been by the Russians and this would have extended the Iron Curtain to Calais. All this may have happened if the Japanese commander had made a different decision at Hawaii!

However, in the final analysis, the ignominious American defeat at Pearl Harbour was the ultimate price to pay for being unprepared!

~NINE~
THE CHANNEL DASH

In February 1942, the German battle cruisers Scharnhorst and Gneisenau together with the heavy cruiser Prinz Eugen slipped out of Brest harbour on the French Atlantic coast. Under Hitler's orders they were embarking on a daring attempt to escape to the safety of German home waters. In so doing the Germans would strike a humiliating blow at British air and sea power by impudently sailing their big capital ships in broad daylight through the English Channel right under the very noses of the British admiralty.

This remarkable achievement by the German navy would succeed where the Spanish Armada four centuries previously had failed and it would leave the British people stunned and infuriated. What was the reason for this audacious Second World War operation? Possibly, there were several factors which influenced Hitler to order the German fleet home. One reason was his concern about the intensification of British air raids on Brest Harbour. In eight months the RAF had dropped four thousand tons of bombs and it was becoming increasingly difficult to protect the ships from this aerial bombardment. There were also the numerous British commando raids on the Norwegian coastline that convinced Hitler that Norway was the most vulnerable region of his Atlantic Wall.

However, possibly the main reason was that he was particularly concerned about the effect the British arctic convoys were having on his war against Russia. These convoys were shipping crucial war supplies around the Norwegian coast to his Soviet enemy on the

Eastern Front. The impatient Hitler was no longer prepared to tolerate three great ships bottled up in Brest; especially when the German Navy was at this time fighting a war on three fronts; namely, the Atlantic, the Mediterranean, and the Arctic. Hitler wanted to concentrate all his surface naval strength in the north, stating that any surface ship that was not in Norway was in the wrong place.

He pressurised his navy chiefs by ordering them to return the capital ships to home waters or they would be decommissioned; and their guns mounted as shore batteries. The Battle of the Atlantic which was now at its height since the Americans had come into the war was to be left to the U-boats alone. This proposal, one should imagine, placed the German navy commanders in a terrible dilemma. On the one hand there was the prospect of the RAF's destructive air power, and on the other the Führer was threatening that unless the ships were brought home he would dismantle them.

The prospect of scrapping the fleet horrified the German navy chiefs as it would present the British with a bloodless victory, and be a death sentence to the German Navy. Admiral Raeder was determined that this would not happen. Most German naval planners had also come to realise that they had few options other than to return their ships through the Channel. To take them by the northern route round Scotland, which had proved a grave for the Bismarck, would subject them to enemy air power operating from land plus the superior Anglo-American battle fleets. Hitler accepted that the chances of bringing the ships safely through the Channel were only fifty-fifty, but, as always, being prepared to gamble, he insisted that a breakthrough was the only viable alternative. However, he was adamant that the risk would only be feasible if absolute secrecy was maintained and providing the enemy was taken completely by surprise.

To the complete amazement of his navy commanders, he ordered the ships to be brought through the Dover Straits in broad daylight. A naval operation of this magnitude within sight of the English coast was without precedent and would be in utter defiance of the traditional use of capital ships. However, Hitler's proposal deserves closer examination; initially it must have appeared tempting for the Germans to sail through the Straits of Dover at night under the cover of darkness. But the disadvantage was that the ships would have to leave Brest in broad daylight, where they were sure to be sighted. If the Royal Navy were to be alarmed immediately, it would give them sufficient time to mass a battle fleet capable of annihilating the Germans. From this viewpoint, it was considered more favourable to go through the Straits in daylight aided by the protective cover of the Luftwaffe.

Hitler persuaded his naval commanders that it would be impossible for the British in the space of just a few hours to get together the forces necessary for an organised attack. He was certain that the British would consider it inconceivable for the Germans to attempt a break out through the Channel in daylight; in this assessment he was absolutely correct. British intelligence, being aware that a break-out was possible, based their interception plan on the assumption that the Germans would leave Brest in day time. This would allow them to approach the narrowest part of the Channel, that is, the Straits of Dover, which is only twenty one miles wide, during darkness. RAF Coastal Command was, therefore, ordered to establish dusk to dawn reconnaissance patrols along the Channel; the navy also ordered submarine patrols during daylight hours outside Brest Harbour.

With the Royal Navy's capital ships far distant at Scapa Flow, the Germans calculated that if surprise was achieved, they could expect opposition from only light surface forces. The main threat was expected to come from air attacks and mine fields. Another obstacle for the Germans was the British radar stations strung along the Channel coast. The Germans knew that the British were ahead of them in radar development. But they were not certain how far ahead, and made plans accordingly; the effect of which will be seen later. Taking all these factors into consideration, the Germans came to the decision that, rather than dismantle the great ships; they would risk fighting them through the English Channel in daylight. The operation was code-named; operation "Cerberus" which Hitler claimed would be the greatest naval exploit of the war. This turned out to be an accurate prediction as the venture was later to be acknowledged by friend and foe alike as an “outstanding and daring combined operation by German naval and air forces”.

The planned schedule for the operation was;

*Put to sea from Brest during darkness,
*Pass through the narrows of Dover - Calais by noon the following day,
*Navigate along the Dutch coast that afternoon,
*Enter the North Sea and arrive at home port in the evening,
*In order to achieve this, it would be essential to maintain a cruising speed of 28 knots.

The British First Sea Lord, Sir Dudley Pound, was at this time sixty five years old, and a sick man. He shared the view that if the Germans decided to break out through the Channel they would reach the Straits of Dover at night; to do otherwise, he stated would be pushing audacity to the point of folly. As the Germans had predicted Sir Dudley was not prepared to risk the capital ships based at Scapa Flow; instead, six Swordfish torpedo carrying planes were moved to Manston on the Kent coast. Nine motor torpedo boats were stationed at Ramsgate; in addition, the admiralty alerted six 20 year old destroyers at Harwich to be prepared to intercept the German ships if they came.

In hindsight, it may be easy to dismiss the behaviour of the First Sea Lord as ineffectual and puzzling. But bear in mind; this was the blackest period of the war for Britain. Rommel's Afrika Corps had come to the aid of Italy in North Africa which had the effect of reversing all the previous year's British gains; and the catastrophe at Singapore was looming. Sir Dudley had still not recovered from the loss of the "Prince of Wales" and "Repulse"; British naval power from Singapore to Scapa Flow was stretched to the limit. Not unlike Jellicoe at Jutland in the First World War, Sir Dudley Pound was "the man who could lose the war in an afternoon". This weighed heavily on him and he was emphatic that, "on no account were capital ships to be brought within range of enemy air attacks". His fear was perhaps justified, for instance, what if several British battleships, operating near the occupied coast of Europe were sunk, or put out of action by determined enemy air attacks; this would undoubtedly change the whole picture of the war, particularly in European waters. In any case, Sir Dudley refused to even contemplate that the

Germans would be foolhardy enough to try and bring their ships through the Dover Straits in daylight. He was convinced they would not be exposed to the RAF who would surely sink them as had the Japanese to "Repulse" and "Prince of Wales".

Strange as it may seem, it did not seem to occur to the admiralty that although the RAF did have a most formidable bomber force whose primary objective was to bomb targets at night from a very great height; they had not been trained for the precision attacks on fast moving ships steaming at thirty knots whilst taking evasive action. The only planes which really had much chance of damaging the big ships were torpedo carriers. The British had two types of these; the Swordfish and the Beaufort. The Swordfish had been used successfully against the "Bismarck", but there were only six available at the channel coast and their inexperienced crews were undergoing training. The leader of this half squadron was Eugene Esmonde, who had distinguished himself in the Bismarck operation. He was ordered to prepare his crew for a night attack on the enemy ships in the event of a break-out from Brest.

These single-engine Swordfish biplanes were designed like First World War aircraft. They had a fabric fuselage, open cockpits, and carried a crew of three, consisting of a pilot, an observer and a rear gunner. Yet these old planes could absorb tremendous punishment; anti-aircraft shells passed through the fabric instead of exploding against it. At Taranto Harbour on November 1940; operating under cover of darkness, carrier based swordfish had crippled and put out of action the main Italian fleet. However, the "stringbags" as they were nicknamed had a top speed of only ninety mph, which made them sitting ducks in daylight.

The other torpedo bomber, the Beauforts, which were possibly the most serious threat to the German ships were not placed on alert. None were moved towards Dover, and this decision would bear heavily on the outcome of the battle. These aircraft were much more effective and more capable to sink or disable the German ships. They could fly at twice the speed of the swordfish, and three squadrons were available. One stationed at Leuchers in Scotland preparing for operations against the "Tirpitz"; the others were based in the south of England near Portsmouth. Therefore, to contest the passage of the German fleet speeding up the Channel, the Royal Navy had assembled six Swordfish torpedo planes, nine MTB's and six 20 year old destroyers. It was a pathetic force to put against a mighty German battle squadron; protected by the greatest air umbrella any ships had ever had.

The Admiralty however, was content that the combined torpedo attacks by the MTB's and Swordfish would slow down and cripple the ships. This would be accomplished while they were within range of the Dover coastal batteries, which would deliver the coup-de-grace. In the event of them managing to slip through the straits, they were to be finished by the RAF and by the destroyers. It was not much of a plan, even for a night attack!

Now that the preparations of each side have been evaluated; it is interesting to assess the actual operation and outcome. On February 11th at 22 15, the Scharnhorst, Gneisenau and Prinz Eugen weighed anchor and slipped quietly out of Brest. At this moment the Germans had their first taste of good fortune; the RAF Coastal Command aircraft on patrol outside Brest harbour experienced a radar breakdown during their departure and, therefore, was unaware that the German battle squadron had sailed. The British submarine on patrol outside the harbour had abandoned watch when light faded, as the admiralty was insistent that the Germans would leave during daylight. A similar mishap befell the other aircraft which was patrolling the outer entrances. This aircraft also returned to base for radar repairs, which turned out to be a blown fuse. Another aircraft was ordered to take over, but was unable to start due to a damp spark plug. By the time these air patrols were resumed the German squadron had left harbour and were already speeding at 27 knots through the English Channel. Worse still, and this was inexcusable, the RAF had failed to notify the admiralty that the Brest patrols had not taken place. This combination of bad luck and inefficiency would emerge as a pattern for the British.

A heavy mist now descended over the Channel causing the other British patrols to be cancelled earlier than usual. Had these patrols been maintained, their radar would certainly have identified the German ships steaming off the French coast. Therefore, before the British woke up to the fact that the big ships were out; the

Scharnhorst, Gneisenau, and Prinz Eugen had won a three hundred mile start. They were escorted by six destroyers, ten torpedo boats, numerous "E" boats, and an air umbrella consisting of three hundred day and night fighters. The German ships racing up the Channel were almost into the Dover Straits, the twenty mile wide maximum danger zone; at this point they would be clearly visible by telescope to observers on the cliffs of Dover. Here they would be confronted with weapons which might prove formidable or even decisive; the big Dover coastal guns. Just how effective were these weapons? After Dunkirk, Churchill, fearing invasion had ordered heavy guns to be mounted on the Dover cliffs. Four 13 inch First World War guns from the old "Iron Duke" class of battleship were found in Nottingham covered in cobwebs and rusting. They were cleaned up and installed near Dover. These clumsy, slow-firing guns were actually quite useless. They were difficult to load, only capable of firing at five minute intervals, and after eighty rounds, the barrel wore out.

Shortly after 10h00, two Spitfires on patrol over the Channel spotted German fighter planes and climbed to engage; they had unwittingly stumbled into the outer guard of the air umbrella protecting the German ships. Almost immediately, flak shells burst around them and they were attacked by a dozen Messerschmitt's. As the Spitfires dodged the enemy fighters they sighted the big ships. The RAF orders were to maintain radio silence over the Channel and to report any sightings after returning to base. But when the Spitfire pilot realised that such a large number of ships sailing so close to the English coast must be the German battle squadron escaping from Brest; he decided this was the occasion to disobey orders, break radio silence, and report to base. At first the report was not taken seriously, nobody wanted to believe that the Germans would be making a getaway through the Channel in broad daylight. This was utterly inconceivable; the British admiralty was totally unprepared and their reaction was to be piecemeal and uncoordinated.

On this particular morning the British radar observers had something else to worry about; the previous night every radar set on the English coast had ceased to function. For the first time the Germans were operating their jamming stations in a big way and not a single British radar instrument was able to function. The ruse had succeeded, and the German squadron had been able to continue on its way undiscovered until the time of the Spitfire sighting. The controller on duty at Biggin Hill interpreted the Spitfire report accurately; at long last, the alarm was raised. The German ships had been at sea undetected for eleven hours. The controller then tried to alert headquarters only to discover the post office telephone lines were down. Eventually, the air officer in command, Air Vice-Marshall Trafford Leigh-Mallory was tracked down; he was in the process of reviewing Belgian air force units on parade and his staff officers refused to interrupt him. They were still unconvinced that it was the German fleet in the Channel, decided that it must be fishing boats and were not prepared to bother their commanding officer.

At 11 05 the Squadron Leader at Biggin Hill insisted that the Beaufort squadrons should be alerted, but he had no authority to do so. He did, however, telephone Esmonde at Manston, who did put his six Swordfish on readiness and requested fighter protection. This was the first decisive step taken by the British armed forces to intercept the German battle squadron as they entered the narrowest part of the Channel, and it was taken by a relatively junior officer. Leigh - Mallory, eventually came to the phone, extremely upset at being disturbed by a mere Squadron Leader. He was extremely rank conscious, but after being given the information he decided that it must be the German ships in the Channel and gave the orders to execute the operational plan. There was in fact a British battle plan code named "Operation Fuller", but the intelligence officer in charge of the operation had locked the plans in a safe and gone on leave and no one could find the key. An hour had now lapsed since the Spitfire's urgent report.

At 12 15, exactly according to their timetable, the Germans arrived at the narrowest part of the Channel; between Dover and the French coast, where they were required to reduce speed and negotiate a mine field. They were at their most vulnerable and feared that the British would be waiting to sink them with massive air and sea attacks and coastal gun barrages. The British, however, had still not recovered from their initial surprise; when they eventually did, the big guns of the coastal batteries opened up. This was after the German ships were through the straits and almost out of range. The German crews saw flashes from the cliffs and several splashes in the sea. Although the shells fell short and uneven, it meant that there was no longer any doubt that they had been detected. Smoke shells were ordered to be fired from the escorting "E" Boats to shield the big ships. The gunfire from the English shore soon ceased when the German coastal batteries in France opened up in reply; no hits were reported.

As the German ships reached the Belgian coastline, the British launched the MTB's who raced out towards them at a top speed of 27 knots. The German ships were now cruising at their maximum speed of 30 mph and the protecting screen of "E" Boats could make 35 knots. The MTB's were out-performed and lighter-armed than the "E" Boats as they came under a tremendous concentration of fire from sea and air. They were forced to fire their torpedoes from maximum range which gave the German ships plenty of time to manoeuvre

The first battle of the straits had been lost. In the meantime Esmonde awaited orders to launch an attack with his Swordfish trained for night operations. To send these slow-moving planes in daylight against the ferocious flak and heavy fighter escorts was certain death. The indecisive senior officers at the admiralty let Esmonde decide whether to attack or not, no one was prepared to give him a direct order on a suicide mission. Esmonde, being a dedicated regular officer, and obviously a very brave individual, felt he had no other choice but to attack. He realised that as the German ships were moving so fast he would need to make an immediate decision. Having a top speed of only 90 knots, if they did not take off at once they would lose the enemy ships. He was told there would be plenty of air cover from the spitfires with whom they would rendezvous once airborne. All six planes lumbered into the air. Without hesitation, the Swordfish crews flew courageously, going to their doom in a futile operation; few would survive this mission. Four minutes after the arranged rendezvous time with the escorting Spitfires the Swordfish were still orbiting over the coast of Kent. Only ten Spitfires had arrived, Esmonde decided that it was now or never. The rest of the escort fighters would catch them up; they never did due to administrative bungling and eighteen young men in six slow, old aircraft went on their way. If any mission could be described as suicidal, this certainly was one.

Three of the Swordfish actually managed to get through the German fighter screen and massive flak barrage to drop their torpedoes on target that were all evaded by the Germans. One after another they were blown to pieces and massacred. There were only five survivors; Esmonde was not one of them. He was later awarded the posthumous VC. As the last of the six torpedo planes blew up and splashed into the sea, the German fighters gave a victory roll over the ships before resuming their patrol. It was 12h45 and the German ships were heading up the coast of Holland into the North Sea. While Esmonde's Swordfish were giving the Germans their first real taste of opposition and being shot to pieces, through a combination of inefficient ground work and bad weather the rest of the RAF fighters either took off too late or got lost in the mist.

The six old destroyers from Harwich now launched their attack; they were also slower than the German ships and had to risk going through a minefield to intercept them. Meanwhile, the torpedo-carrying Beauforts, the RAF's most important ship-busting aircraft were finally ordered to fly to Manston. Delayed by ground staff incompetence; some of the Beauforts were armed with bombs and not torpedo's. The distance from their base in Portsmouth to Manston where Esmonde's Swordfish were based is one hundred and twenty miles; if they had left at the same time as Esmonde took off they could, by flying at twice his speed, have joined him to deliver a coordinated attack. Inexplicably, rather than let those armed with torpedoes take off and join Esmonde it was decided to hold back the entire squadron until they were all ready. The Beauforts from Leuchers in Scotland arrived over Manston and circled for over an hour awaiting orders which Manston ground staff were frantically signalling on the wrong frequency. No one had told Manston that these aircraft had recently had their radio frequencies changed. It was only when they were forced to land and re-fuel that this botch-up was discovered. In addition to this; several of the Beauforts landed at Manston expecting to be armed with torpedoes, only to find that none were available. No one had checked how many were in store.

When they eventually set off, it was with orders to attack a German merchant convoy. No one had told them of their real mission, and in the utter chaos, the Beauforts' Spitfire escorts flew off in another direction looking for them. At 3 45 pm. a few of the Beauforts found the German ships, but due to the lack of fighter protection and the concentration of flak; dropped their torpedoes at maximum range, missed the target and returned to base. While the torpedo-carrying Beauforts were struggling against the continuing ground muddle, Bomber Command finally came to life. Six hundred and seventy five aircraft took off, most of which had been on bombing raids over Germany the previous night. The majority of the aircraft could not find any target and did not really know what they were looking for. At one stage a squadron of Hudsons was told to follow the Beauforts out to the German ships. The Beauforts, however had been told that they were to follow the Hudsons to the target; the result was that every time the Hudsons formed behind the Beauforts, the Beauforts would fall back behind the Hudsons. This ring-a-ring of rose's ridiculous situation continued for half an hour before the Wing Commander of the Beauforts became totally exasperated and decided he had enough and led his aircraft out to sea.

The British destroyers by this time had worked their way through the minefield, but never got within striking distance of the German ships. In many instances they were attacked indiscriminately by both British and German aircraft. They did, however, press forward to try and launch their torpedoes from a range of two miles. All missed and the destroyers, though severely damaged managed to limp back to port.

The German ships succeeded in avoiding the bombs and shells by skilful manoeuvring. Mines, however, proved to be a greater danger; at 14 31, the leading ship, Scharnhorst hit her first mine off the Dutch coast and was lifted bodily out of the water. Electric installations failed, the rudder no longer functioned and the gyro-compass was out of action. The Scharnhorst was left rolling helplessly in the North Sea swell; a sitting target for the vengeance of the British. The Scharnhorst crew worked feverishly to repair the damage whilst the Luftwaffe and escorting vessels prepared to hold off the enemy who never came. The ships engineer's surpassed themselves and in just over half an hour got the ship going again. The Scharnhorst, however, hit a second mine in the evening as she approached German waters. With one thousand tons of water in her hull, and almost out of control, she nevertheless reached port due to the coolness and excellent seamanship of her captain. Later that same evening, Gneisenau struck a mine, but also managed to reach port.

What were the result and the reaction to this operation? The German losses were seventeen Luftwaffe planes, two torpedo boats damaged by bombs, two dead and several German sailors wounded. All Germany rejoiced over the feat which was hailed as a mighty blow against an enemy who had ruled the waves for centuries. At 1 00 am in London the First Sea Lord, Sir Dudley Pound conceded that the German ships must have reached the safety of home waters. He phoned Churchill to make one of the worst reports a British admiral has ever had to make to a British prime minister. Typically, Churchill growled "why" and slammed the phone down. The RAF officers blamed the disaster upon the fact that very few pilots really knew what they were looking for due to the ridiculous lack of information. One totally frustrated Beaufort Squadron Leader declared, *"I was sent looking for a convoy, why was I not told about the bloody great battleships"?* He continued, *"The reasons the German ships got through is not that the plan to stop them was badly executed or that it miscarried; the real trouble is that there was no plan at all. We relied far too much on hasty improvisation".*

Nearly seven hundred fighters and bombers had been flung into the battle without success because they were too late and completely uncoordinated. Thirteen young fleet air arm pilots had been sent uselessly to their deaths; twenty seven young sailors had been killed and eighteen wounded. The fury unleashed by the military commanders was not isolated; it sent an electric tremor of outrage through the country. The nation was ashamed and affronted; British minds went back to Sir Francis Drake and the Spanish Armada; in effect, it spelled the end of the Royal Navy legend, which decreed: *"in wartime no hostile warships would be permitted to neither approach so close to the English coast nor pass through what was proudly called the "English Channel".*

In the entire British press a storm of indignation arose; how could this have happened right under their very noses? British public opinion was furious at the success of the channel dash. The war cabinet was violently attacked, causing a crisis of confidence in Churchill's wartime cabinet. The times, for example, thundered, *"the German navy has succeeded where the Spanish Armada of 1588 failed.*" It continued, *"nothing more mortifying to the pride of our sea power has ever happened in home waters".*

National and provincial newspapers joined with the times in voicing indignation that this humiliating disaster had been allowed to happen. The news chronicle stated *"though individual courage and steadfast devotion to duty was evident, those primarily responsible for the war's greatest blunder must be brought to book".* The newspaper attacked Churchill's war time cabinet by stating: *"the incident is symptomatic of a general feeling that something is wrong with Britain's war effort; the tired and incompetent people occupying high office should be removed".*

Conversely, while British newspapers roared their disapproval, the Germans were ecstatic with their victory. Especially the reference to the Spanish Armada; Hitler had been proved right. Faced with the same allegation of "bungling in high office" which forced him out of office in the First World War; Churchill took the unprecedented step, during war time that is of ordering a military tribunal to be set up to investigate the break-out. The findings, however, of this report were shrouded in secrecy and not even the members of parliament were allowed to see it. Churchill as defiant and truculent as ever, scowled at questions in the house by saying *"our affairs are not conducted entirely by simpletons and dunderheads, as the comic papers try to depict".*

Despite this attempt at reassuring the nation, there was much evidence of reports concerning the "Channel Dash" of going missing and officers being shunted to lesser posts. Three days later on February 15th, the British army surrendered at Singapore in one of the most ignominious fiascos of modern military history. The "Channel Dash" plus Singapore made black February of 1942 the lowest point of the entire war for Great Britain. Churchill, now sixty seven years old was under severe pressure to throw in his hand. However, the pugnacious war time leader was not prepared to surrender high office. In retrospect, perhaps it is understandable that Churchill should cover up the disaster. The British public were already despondent enough about defeats from Dunkirk to Singapore. There was no point in dismaying them further by revealing the full facts of the military and bureaucratic inefficiency which allowed the German warships to sail unscathed past the cliffs of Dover. The unpalatable truth which Churchill dared not reveal to the angry and disturbed British public was that some of his service chiefs had proved themselves to be tragically incompetent. It makes one wonder what might have been the outcome if Hitler had gone ahead with his plans to invade Britain.

The break-out of the ships was a supreme example of meticulous German planning and efficiency; defeating the hasty last minute improvisations of the British, however, there were compensating advantages for the British. The major one being that the threat from Brest to the Atlantic convoys had been eliminated and no further attempt would be made to throw German capital ships into the battle of the Atlantic. Churchill stated after the war *"viewed in the aftermath and in the overall context, the episode was of great benefit to us"*. One man who agreed with him was Grand Admiral Raeder, who during interrogation at the war's end, commented *"it was a tactical success but a strategic defeat"*.

These points merit closer analysis; why was the channel dash not a total defeat for Britain? Perhaps the answer to this question is that although the German ships achieved startling success in escaping through the Channel, their careers as fighting ships soon ended. A fortnight after arriving back in Germany, Allied bombers had their revenge by finishing off the "Gneisenau". The damage caused by the mine struck whilst escaping up the Channel had made it necessary for her to enter the dry dock at Kiel for repairs. The RAF took full advantage of this opportunity and made her the target of massive air attacks; smashing her bows and foredecks. It was the end of the "Gneisenau" as an operational sea fighting ship. She was towed to Danzig, her hulk was filled with concrete and she suffered the ultimate humiliation of becoming a block ship.

After repairs, the "Scharnhorst', only got into Norwegian waters in March 1943. Her fate was sealed in December 1943 when she was cornered off the North Cape by the British home fleet. The "Scharnhorst" was put out of action by the battleship "Duke of York" and the escorting destroyers finished her off with their torpedoes; only thirty six of Scharnhorst's crew, who had participated in the Channel Dash, out of a complement of almost two thousand survived.

"Prinz Eugen" unscathed in the Channel Dash, was soon ordered to make for Trondheim in Norway. But was torpedoed en route by a British submarine and had to turn back, never venturing to sea operationally again. At the war's end the Americans took possession of her and she was sunk in atomic bomb tests

In the final analysis it must be realised that the German navy's brilliant exploit, and it should be it remembered was one of Hitler's better inspirations was of no great long term military significance. Nevertheless, the Channel Dash was a major triumph for the Germans. For the British, it was a terrible humiliation, and very nearly brought down the Churchill administration, with all its ramifications.

~TEN~
THE FALL OF SINGAPORE

The Battle for Singapore commenced when Japan invaded Malaya on the 8th December 1941, the day after Pearl Harbour. Improvements to the defences of Singapore had only just been completed at great cost. Singapore was considered a strategically vital military base, and thought to be an impregnable fortress; frequently referred to as "The Gibraltar of the East" and "Bastion of the Empire."

The fall of Singapore was the greatest humiliating defeat in the history of British arms. One hundred and thirty thousand Commonwealth troops surrendered to thirty thousand Japanese. Winston Churchill called it the "worst disaster" and "largest capitulation" in British history.

Commonwealth forces, mainly comprising of British, Australian, and Indian troops were inadequately trained and inappropriately deployed; mistakenly believing that the Malay jungle was impassable. Therefore, the Japanese onslaught when it came through the Malay Peninsula took the British Command by complete surprise. The British military command in Singapore did anticipate an attack from the sea and were confident it would be easily repulsed. This explains why all the guns at Singapore pointed out to sea.

Commonwealth troops were told that the Japanese were poor

fighters and not a threat to the mighty British Empire. They had become victims of their own propaganda. An air of complacency and lack of preparation was evident. Social life was considered more important in Singapore. The Raffles Hotel and Singapore Club were often frequented by officers. To the British, war was still fought by the 'rule book. However, this was soon shattered. The Japanese had no intention of fighting a conventional form of war. With speed and savagery, using bicycles and light tanks for swift movement, they never allowed time for defenders to re-group. They were ordered not to take prisoners as this would slow down the advance. Some captured troops were doused with petrol and burned to death. Locals who had helped the British were tortured and murdered. The brutality of the Japanese soldiers was horrific.

Britain's naval presence stationed at Singapore consisted of a squadron of warships named force Z; led by the modern battleship "Prince of Wales" plus the battle cruiser "Repulse." This force under the command of an Admiral resolutely disbelieving in air power was meant to be a deterrent to the Japanese.

On December 8th 1941, the squadron put out to sea and headed north up the Malay coast to where the Japanese were reported to be landing. On December 10th, both ships were sunk by air attacks from Japanese land based bombers and torpedo bombers; thus, making history by being the first capital ships to be sunk in open seas from the air. The loss of both capital ships had a devastating impact on morale in Britain. Winston Churchill wrote in his memoirs: *"In all the war I never received a more direct shock"*.

The Japanese were led by General Yamashita, known to his troops as "the tiger of Malaya." The Commonwealth troops were led by General Percival, nicknamed "the rabbit" by his troops. Both men were probably appropriately named. Many of the Japanese troops had fought in the Chinese campaign and were battle hardened; whereas many of Percival's men had never seen combat.

Singapore's air defence was initially provided by only outdated Brewster Buffaloes. So with the Mitsubishi Zero, the Japanese soon gained air superiority. The ground defence forces had no tanks and few armoured vehicles, which put them at a severe disadvantage.

On January 31st 1942, the Commonwealth forces withdrew across the causeway that separated Singapore from Malaya at the Johore Strait. The causeway was then demolished. Percival spread his men across a seventy mile line. This proved a mistake; he had overestimated the strength of Yamashita's forces. Japanese infiltrators, many disguised as civilians, crossed the Straits in inflatable boats and reported back intelligence information. From aerial reconnaissance Yamashita and his staff gained excellent knowledge of the defenders positions.

On February 8th the Japanese attacked across the Strait. The superiority of artillery, aircraft and military intelligence, tore gaps in the thinly spread defence lines. Japanese tanks were towed across the strait and advanced rapidly. Yamashita outflanked and bypassed the defenders, forcing them to retreat.

Singapore's coastal guns were supplied mostly with armour piercing shells and few high explosive shells. Armour piercing shells are for penetrating the hulls of warships and are ineffective against land personnel. Most of the guns could be directed northwards and did fire at the invaders. If the guns had been well supplied with HE shells the Japanese would have sustained heavy casualties.

Due to an inadequate early warning system, Singapore city was subjected to intense aerial bombing. The bombers struck the city centre as well as the naval base and air fields, causing many civilian casualties.

On the evening of 10th February, British Prime Minister Winston Churchill, cabled General Wavell, the British commander in the region, stating: *"There must be no thought of saving the troops or sparing the population. The battle must be fought to the bitter end at all costs. Commanders and senior officers should die with their troops. I rely on you to show no mercy to weakness in any form. With the Russians fighting desperately for survival and the American stubborn resistance in the Philippines; the honour of the British Empire is at stake".* Wavell subsequently ordered Percival to fight on to the end, and that there should be no surrender in Singapore.

On 11th February, aware that his supplies were running perilously low, Yamashita decided to bluff, committed his reserves and called on Percival to *"give up this meaningless resistance"*. The Japanese had captured most of the ammunition and fuel and controlled the main water supplies. The defending troops tried to hold the Japanese at various defensive lines. But on 13th February, they were forced to retreat further. Senior officers advised Percival to surrender in the interests of minimising civilian casualties. Percival refused.

The Japanese were now within five kilometres of Singapore and the entire city was now within range of Japanese artillery. Bombing and artillery fire intensified and civilian casualties steadily mounted. On 14th February the Japanese captured the main water supply reservoir and advanced towards Alexandra Barracks Hospital. A British lieutenant, acting as an envoy with a white flag, approached the Japanese. He was bayoneted and killed. When the Japanese troops entered the hospital, a number of patients, including those undergoing surgery, were killed along with doctors and nursing staff. The Japanese did not consider those who were wounded or who had surrendered to be worthy of survival.

By the morning of 15th February, the Japanese had broken through the last line of defence where troops were running out of food and ammunition. The anti-aircraft guns had also run out of ammunition. The bombing, fighting and heavy shelling continued. The hospitals were crowded and overflowing. Many of the troops became separated from their units.

At 09:30, Percival held a conference with his senior commanders. He proposed two options. Either launch an immediate counter-attack to regain the reservoirs and the military food depots or capitulate. All present agreed that no counter-attack was possible. Percival opted for surrender. A deputation was selected to go to the Japanese Headquarters. They set off bearing a Union Jack and a white flag of truce towards the enemy lines.

They returned with orders that Percival himself proceed with Staff Officers to the Ford Motor Factory, where General Yamashita would lay down the terms of surrender. Percival issued orders for all secret documents to be destroyed and formally surrendered shortly after 17.15.

The terms of the surrender included:

> *The Japanese Rising Sun Flag to be hoisted over the tallest building in Singapore.
> *The unconditional surrender of all military forces in Singapore.
> *All troops to remain in position until further orders.
> *All weapons and military equipment to be handed over intact.
> *To prevent looting, etc., a force of one thousand British armed troops to take over until relieved by the Japanese. Yamashita accepted full responsibility for the lives of all servicemen and civilians remaining in Singapore.

Despite the instruction to Australian troops to stay at their posts, General Bennett and two of his staff officers commandeered a small boat on the night of the surrender. Eventually, and with much controversy, making their way back to Australia. General Bennett later said*: "The whole operation seemed incredible: five hundred and fifty miles in fifty five days – forced back by a small Japanese army riding on stolen bicycles".*

The Japanese occupation of Singapore had begun and Japanese newspapers triumphantly declared the victory. Singapore residents would suffer great hardships during occupation and fifty thousand people of Chinese origin were slaughtered. Many of the Commonwealth soldiers taken prisoner would never return home. Thousands were shipped on "Hellships" to other parts of Asia to be used as forced labour on projects such as the Burma Death Railway.

The humiliation of Great Britain at the hands of a numerically smaller Asiatic army was to have fateful consequences for the Far East. The Singapore catastrophe spelt the death knell of the British Empire. Today, debate still rages about why the defence forces in Singapore were so comprehensively beaten.

When Lieut.-General Percival ordered his troops to lay down their arms on 15th February 1942, the Pacific War was just ten weeks old.

In that time, Japanese forces had steadily driven the forces defending the Malay Peninsula southwards in a relentless but short campaign. The Commonwealth ground troops were simply outclassed by the Japanese. The very speed and apparent ease of the Japanese victory had a severe impact in both Britain and Australia.

Singapore had stood for many years as a potent symbol of British power in South-East Asia. Yet within weeks of its surrender the Japanese were bombing the Australian port of Darwin. Australians discovered too late that the fundamentals of the defence policy were unsound. Britain had promised to provide a fleet for the base to deter Japanese aggression. But, when that situation finally arose, the Royal Navy was already heavily committed in the Atlantic and Mediterranean.

What arrived early in December was not a great fleet but a small squadron based around just two capital ships that were quickly disposed of in the opening days of the Japanese invasion. Singapore thus remained without the fleet that was considered its primary means of defence. Military planning for the air defence had called for a minimum of five hundred aircraft. But this requirement had never been met, either in the number of aircraft provided or effective types. Despite their best efforts, Malaya's aerial defenders were at the mercy of an enemy that was vastly superior in air power.

It has been claimed in a contemporary British report released in 1992 that it was acts of indiscipline by Commonwealth troops in chaotic circumstances which undermined the defence and directly contributed to the surrender. Acts of indiscipline there almost certainly were, considering the underlying weakness in defensive arrangements. However, there can be no doubt as to where the real cause lay for the loss of Singapore.

Churchill is very severely taken to task. He had full knowledge of the deficiency in aircraft at Singapore. Nevertheless he accepted the loss of over two hundred planes in Greece, and then sent towards the end of 1941 another six hundred to the Soviet Union. Had these aircraft been sent to Singapore, the air defence in Malaya would have consisted of a total of eight hundred modern aircraft instead of one hundred and forty old-crocks.

The failure to give due recognition to the technical capacity of the Japanese air forces was not only Churchill's but that of the highest ranking officers in the R.A.F. Furthermore, it could be argued that by sending Stalin hundreds of fighter planes, the British Premier acted in the conviction that he was defending Britain by keeping Russia in the war. To him, the real danger towards the end of 1941 was the collapse of the Red Army. Hitler would then turn his full power against Great Britain. Her chances of survival would have been slender, and Singapore would inevitably follow her into defeat.

Perhaps no aspect arising from the fall of Singapore was more galling than the monstrous fate experienced by the troops taken prisoner. These men, and a small number of Army nurses, were subjected to brutal treatment at the hands of their captors. Those sent on work gangs for the Burma Railway especially suffered from overwork, malnutrition, sickness and beatings from guards. More than a third did not survive the war.

After the Japanese surrender in 1945, Yamashita was tried by a U.S. Military Commission for War Crimes, committed by Japanese personnel in the Philippines. But not for crimes committed by his troops in Malaya, Burma or Singapore. He was convicted and hanged in the Philippines on 23rd February 1946.

~ELEVEN~
THE RAID ON St. NAZAIRE

The raid on St. Nazaire in March 1942 was one of the most outstanding acts of heroism in World War Two. More VC's were won during this classic combined services raid than in any other operation in the war.

It should be remembered that at this time the U-boat offensive was at its height. Therefore, the deployment of a powerful enemy surface raider into the Atlantic might have proved decisive. The St. Nazaire dock was the largest dry dock in the world, able to accommodate a ship of eighty five thousand tons and contained a fully-equipped submarine base.

The Tirpitz, then in Norwegian waters was the most powerful battleship in the world, and St Nazaire, where she could be maintained and repaired, was the key to her wider operations in the Atlantic.

Because of the enormous size and the great pressure of water that had to be handled in the dry dock; it was closed at both ends by four steel sliding lock gates fifty metres long, twenty metres high, and ten metres thick. Each gate was pulled by winding gear housed at the dock. There was also, a pump house, with motors at ground level and pumps ten metres underground. Five thousand German troops guarded the dock area, and another one thousand manned the guns along the banks of the estuary to protect it from sea and air attack.

In January 1942 Lord Louis Mountbatten, Director of Combined Operations, was requested to examine the prospect of an attack on St Nazaire. A plan was developed taking into consideration the latest intelligence information from all sources.

The German defences were extremely strong; armed with numerous dual-purpose 40-mm, 20-mm and formidable 88mm guns. Most of the guns were mounted on the tops of concrete bunkers, flak towers, and on the roofs of the submarine pens. There were also two heavily armed ships mounting several formidable 88-mm. guns.

During February, Lieutenant-Colonel Newman, commanding No. 2 Commando, was ordered to draw up a list of one hundred men to undergo special training. Commander Ryder was appointed naval force commander. The plan was to ram the lock gates with a destroyer and destroy dock installations.

HMS Campbeltown, once the USS Buchanan, was converted for this purpose. Two of her four funnels were removed, and the other two given a rakish German look. Her original armament was replaced by one 12-pounder on the foredeck and eight Oerlikon cannon. The bridge and wheelhouse were armour-plated and her decks cleared except for ten armoured screens, intended to provide shelter for the commando's during the attack. Delayed action explosive charges were placed in the bow. They were then encased in concrete in order to disguise the Campbeltown as a blockship. The ship was to be scuttled after ramming, thus making it impossible for her to be dragged clear of the dock gates before the delayed action explosion.

In addition to commandos aboard the Campbeltown, other teams would travel in motor launches which would have the advantage of easy navigation in the channel into St Nazaire. A motor torpedo-boat, commanded by Sub-Lieutenant Wynn, was given the task of entering the dock and place delayed action torpedoes at the inner gate. A motor gunboat would carry the leaders of the expedition into the attack.

By March 25th the force under the code name 'Chariot' was assembled at Falmouth. Two destroyers were to provide escort until night fall immediately before the attack. At 14h 00 next day, under a clear sky, the ships carrying the commandos sailed out of Falmouth. The total force numbered six hundred and eleven men. During the following afternoon the sky became overcast as the force moved in towards the French coast. By 20h 00s it was dark and time for the escort destroyers to leave. A German ensign now fluttered from each vessel.

At this moment sixty RAF bombers were taking off from England, bound for St Nazaire. Selected targets had been clearly identified. The first wave arrived at 23h 30 and ran into heavy flak. A few bombs were dropped, but cloud cover dropped well below six thousand feet. Unfortunately, the air raid served to arouse the suspicions of the defenders and all Wehrmacht command posts were alerted.

Newman's force entered the Loire estuary at 00h 30. Twice Campbeltown grounded, but she drove on into deeper water. Below her decks the time fuses had been activated. At 01h 20 the gunboat leading the force passed the port entrance. Suddenly, a searchlight flooded the force. From both ahead and to starboard German signal lamps flashed their challenge, and two 20-mm guns sent warning bursts overhead. Leading-Seaman Pike, an expert in German signals, raised an Aldis lamp and began spelling out a long complicated signal claiming that the force was proceeding to harbour in accordance with orders.

The speed of the vessels increased; so did German doubt. Ryder fired a Verey pistol as a recognition signal. This failed to reassure the Germans; it was the wrong colour. Every German gun that could be brought to bear opened fire and every British gun replied. The German ensigns were pulled down from each masthead and replaced by the white ensign.

As the biggest ship, the Campbeltown was immediately the prime target. She took several direct hits and enemy fire took a heavy toll of the tight-packed commandos. Two hundred metres from the dock entrance, the dark bulk of the dock gates appeared ahead; Campbeltown's commander increased speed to 20 knots and gave the order: 'Stand by to ram!' Half the men aboard were dead or wounded, gallantly, with all remaining guns blazing, she crashed into the outer gate.

Six minutes after Campbeltown struck, Newman and his party were ashore, running between the warehouses towards the building he had chosen for his headquarters. In the dock area, came the stutter of machine guns and the sharp explosion of grenades as the party from Campbeltown went into action. The officer in charge of the demolition squads landed and set out with his team to neutralise the locks. Charges were placed under the bridge, but the party then ran straight into German defenders and were cut down.

Gun and searchlight positions were stormed and the weapons destroyed. German gun teams on the roof of the pumping house fled before the kilted Highlanders could reach them. The guns were blown up, and the party dropped back to hold the bridge over the Old Entrance. Lieutenant Chant and four sergeants faced the task of blowing up the pump house. Chant had been wounded in the knee before he left Campbeltown, and one of the sergeants, badly wounded, had to be carried by the others. Nevertheless, in the pump house the charges were placed on the pumps, and the resulting detonation collapsed the floor of the pump house, taking the control machinery with it.

The winding house for the gate in which Campbeltown was embedded had already been wrecked, but at the far end of the dock a bitter battle was raging. The demolition team for the northern gate winding mechanism had laid their charges, but detonation had been delayed because other teams were still laying charges in the area. To make matters even more dangerous, 20-mm guns were now raking the party working on this gate.

At last the main charges were laid. Sergeant Carr, ordered his men clear, and blew the charges. Water poured through the gate into the dock. As he withdrew the winding-house blew up. Demolition in the dock area was now complete.

The river was an expanse of burning petrol, upon which floated the wrecks of motor launches and Wynn's motor torpedo-boat. Some of the boats, including the leading motor launch, had erupted in flame even before Campbeltown struck; their thin skins ripped open by gunfire. Most of those aboard were killed or drowned, the few survivors were captured. Two motor launches did, however, get into the Old Entrance, and the first of these waited to pick up the demolition team before returning to the river.

But searchlights and burning petrol lit the scene, and Campbeltown burst into flame as tracer shells hit the fuel tanks. Another motor launch, took aboard 30 of the Campbeltown's crew and sped back to the river mouth. There Campbeltown was hit by a shell from the coastal batteries, and only three survivors were picked up the following morning by a German trawler. Only one of the motor launches succeeded in landing its commandos, but was caught by concentrated fire as it returned to mid-stream, and blew up.

Ryder, ordered Wynn to use his delayed-action torpedoes against the outer gate, then pick up as many of Campbeltown's crew as he could and make for home. The torpedoes struck home and sank, sixteen of Campbeltown's men were picked up, and the torpedo-boat raced downstream at 40 knots.

In a moment of compassion, Wynn suddenly stopped his engines and went astern to pick up two men on an oil-covered Carley float. At this moment the torpedo boat was struck by a heavy shell and burst into flames. Of the sixteen men aboard, only four survived to reach a prison camp, including Wynn. After Ryder had satisfied himself that Campbeltown was sinking, he and Leading Seaman Pike rejoined the gunboat.

At about 02hrs 30 Ryder moved out into the river so that he might weigh up the general situation. The scene that met his eyes was appalling, indeed catastrophic. Alone on the river except for the dead and the dying, Ryder and his crew headed for the open sea through floating wreckage and bodies.

A German destroyer flotilla patrolling in the area, received a message that enemy forces were landing at St Nazaire. The flotilla turned back. A German destroyer picked out the tiny shape of a motor launch, carrying fourteen crew and fourteen commandos. The German commander called upon the launch to surrender rather than fight a battle which could have only one ending, but hardly had his searchlight focused when fire from Oerlikon, Lewis, and Bren guns poured into his ship and his searchlight was smashed. The German commander replied with heavy machine-gun fire which raked the tiny launch mercilessly, killing and wounding all on board. In the growing daylight the launch drifted helplessly and bodies lay strewn across the deck.

Sergeant Durrant became the first soldier to win the Victoria Cross in a naval action, and also probably the first man to win so high a decoration as the direct result of a recommendation from the enemy.

What remained of the force ashore, as the Germans hunted them down, split into small units and tried to escape out into the open country. Not many were able to get clear of St Nazaire; those caught were herded together in various German centres. One of the survivors was Campbeltown's commander Lieutenant-Commander Beattie, who had been picked up by a German trawler after his launch had been blown up. He was interrogated by a German Intelligence officer who remarked patronisingly:

'Your people obviously didn't know what a hefty thing that lock gate is. It was really useless to try to smash it with a flimsy destroyer'.

At that moment an enormous explosion shattered the windows of the office in which the interrogation was taking place, and a vast black cloud shot up from the Campbeltown. The lock gate disintegrated and a tidal wave flooded in, carrying the shattered destroyer half way along the dock.

'That, I hope,' said Beattie 'is proof that we did not underestimate the strength of the gate'.

The following day, the first of Wynn's delayed action torpedoes blew up; and an hour later the second torpedo exploded. Operation 'Chariot' was over.

Of the eighteen craft which had entered the Loire estuary, only two launches returned home. Of the men who had sailed in them, one hundred and sixty nine had been killed, and only five of those left ashore managed to make their way home by escape routes. Two hundred and fourteen went into prison camps. Their sacrifice was most certainly not in vain: the dock was never used again by the Germans, and the Tirpitz never entered the Atlantic.

Described as "The Greatest Raid of All," it was a huge success. The Tirpitz was effectively trapped in Norwegian coastal waters and was sunk before she could sink a ship herself.

Stephen Beattie, together with four others was awarded the Victoria Cross for his captaincy of the Campbeltown.

The success of the raid gave Britain a major boost in confidence. With Churchill's support, the Commando idea took root.

However, this success, and subsequent Commando raids, led to Hitler's grim 'Commando Order', from then on any member of the Special Forces captured alive was to be shot. The terms of the Geneva Convention would not be extended to Commandos.

Many years afterwards, at a memorial service held on the scene of their triumph, the survivors were told by the French Prime Minister: *'You were the first to bring us hope'*. It was a fitting epitaph recognised by the award of five VC's.

~TWELVE~
BOMBER COMMAND

Fifty five thousand dead, twenty thousand wounded, and twelve thousand taken prisoner. More commissioned officers and NCO's than the British armed forces lost during the entire First World War. That was the sacrifice made by the airmen of RAF Bomber Command in the most protracted and bloodiest campaign of the Second World War.

Were their young lives sacrificed in vain, and did area bombing really shorten the war? Was the bomber offensive a legitimate act of war or was it a crime against humanity? These are some of the issues that continue to be contentious. In any war the rationale that the victors and rarely the vanquished determine ethical values is applied. This was clearly spelled out at the Nuremberg trials.

The Germans started the bombing campaign when they bombed Poland in 1939, and was only ended six years later, when the Americans dropped the atomic bombs on Japan. The bombing of towns and cities became a matter of angry recrimination amongst various groups; who argued long and heatedly over the conduct, the morality and even the effectiveness of area-bombing. Especially as the person at the centre of this controversial and fiercely disputed subject is none other than Sir Arthur Harris the Commander in Chief of RAF Bomber Command from 1942 to 1945.

It was he who developed the "saturation" bombing technique, and with great passion and resolution he set out to demolish Germany from end to end with the intention of breaking the will of the German people. Declaring emphatically that area bombing alone would win the war, he proceeded to convert Bomber Command into an awesome destructive power to carry devastation into the heart of enemy territory, most notably Hamburg and Dresden.

Why did Harris fanatically pursue the destruction of German cities and the slaughter of hundreds of thousands of civilians? Did he perhaps misinterpret the policy of the Allied leaders at the Casablanca Conference, when they issued the following directive to their Air Force Commanders? *"Your primary objective is to deny the enemy the ability to wage war by the systematic destruction and dislocation of the military, the industry and the economy to a point where resistance and morale are fatally weakened".*

There are four issues here to consider, and the question arises; did Harris adhere to all four, namely, military, industrial, economic, and morale. Or, did he focus only on the fourth issue, which is morale and ignore the other issues?

This is what should be attempted to come to terms with, because few issues have remained as disputed as the contribution area bombing made to the Allied victory. It is contentious mainly due to the enormous amounts of material and manpower mobilised by Harris. This expenditure and energy is openly repudiated by some military historians, who claim that the total investment in manpower and resource, by far, surpassed the amount of damage inflicted on the enemy's war effort. They argue that the results achieved bore no relation to the cost; and this material and manpower should have been invested elsewhere.

The nation's energies were devoted to supporting the bombing offensive and a great number of her best and bravest young men gave their lives to carry it out, though in the end it did little to hamper the German war machine, but much to sow the seeds of post war hatred. It also undermined Britain's standing in the world as the upholder of civilised standards of behaviour. On the other hand, Harris claimed that his bomber offensive caused Germany to collapse psychologically and physically. He states that his air offensive made a decisive and indispensable contribution to victory and resulted in the paralysis of Germany's industrial base, leading to her total defeat.

Statistics, however, disagree, German output continued to defy the aerial onslaught until the last months of the war. For example, aircraft and tank production continued to increase until the end of 1944. Therefore, taking into consideration the total tonnage of bombs dropped and the horrendous toll inflicted on Bomber Command, was the return for such prohibitive losses disproportionate? To try and answer this, it is necessary to go back to the strategy adopted by Prime Minister Neville Chamberlain at the outbreak of World War Two in September 1939.

On the forlorn hope that all-out war could still be averted, he issued stringent instructions that non-combatant casualties were to be avoided at all costs. Only naval targets were to be attacked and aircraft were to return to base with their bomb loads rather than drop them near residential areas where civilians could be endangered. Private properties, including industrial sites were to remain undamaged, and the main RAF task was to drop propaganda leaflets over enemy territory. Winston Churchill, then at the admiralty, dismissed this leaflet activity with the wry comment; *"The enemy is hardly likely to be induced to surrender by supplying him with free toilet paper"!*

The events on the afternoon of 18th December highlighted the futility of this ludicrous situation when twenty two RAF Wellington bombers raided Wilhelmshaven but turned away without attacking due to fear of damaging nearby properties. As they withdrew, enemy fighters intercepted them, and only ten severely damaged planes, still with their bomb loads, returned to base. This suicidal restraint came to an abrupt halt when Hitler struck west and the Luftwaffe bombed Rotterdam. Part of the city was reduced to rubble and several thousand civilians killed. There then followed the swift defeat of the French, which presented the Germans with an ideal opportunity to invade Britain and end the war. Fighter Command, however, denied them control of British airspace.

Of even more significance at this time, that is, from a strategic point of view, Bomber Command attacked and destroyed many of the landing barges that were being assembled in occupied Channel ports. Thus averting the threat of invasion and for the first time in history command of the air combined with strategic bombing decisively ended a military campaign. That is "Operation Sealion".

Britain then had to endure the Blitz for almost a year until Hitler attacked Russia. During the Blitz, the indiscriminate bombing of Coventry caused the deaths of five hundred and fifty people and destroyed fifty thousand homes. Many other British cities were also bombed, but London was the principal target and between September 1940 and May 1941, more than seventy major raids were launched on the capital. A total of forty thousand British civilians died during the bombing and approximately one million dwellings were destroyed. But, although it caused a great deal of damage, the Blitz failed to inhibit British production capacity or seriously dent morale. In fact, once over the initial shock, civilians came to take pride in making do and carrying on as normal. The spirit of camaraderie brought people together as never before and the effect was to toughen British resistance and increase their resolve to retaliate by aerial bombardment.

The Germans would eventually pay a terrible price for their indifference to civilian casualties. Their major towns and cities would be subjected to relentless air attacks eventually destroying three and a half million homes and leaving over seven million people homeless.

In early 1942 Air Marshall Arthur "Bomber" Harris took over Bomber Command. He was determined to avenge the destruction of British cities by escalating aerial bombardment, and targeted the largest centres of population, which, he believed would "break the spirit of the German people". He captured the imagination of the British public by gathering all available operational aircraft to attack Cologne in the thousand-bomber raid. This marked the beginning of area bombing, or according to the Germans, "terror-bombing". Improved navigation equipment enabled British bombers to successfully locate and hit their target. In a period of just ninety minutes, eighteen thousand buildings were destroyed, leaving fifty six thousand Cologne citizens homeless. Aircraft losses were less than

4%, which was quite incredible compared with later losses.

Harris was convinced that a concentrated campaign against Hitler's other big cities would bring about the collapse of Germany, the reality, however, was quite different. Harris's objective had been the destruction of the city of Cologne, and although damage was widespread, this did not happen. The city was only seriously disrupted for two weeks. Harris disregarded this evidence and demanded that all Coastal Command and army support aircraft be absorbed into Bomber Command, arguing that the invasion of occupied Europe was unnecessary. The Admiralty, in the midst of the Battle of the Atlantic and struggling to get the convoys through were exasperated by this prospect. In their view Cologne was merely a theatrical enterprise.

There was no military reason for the thousand-bomber raid on Cologne, and they in turn argued that the transfer of Coastal Command aircraft to the bombing campaign over Germany could cost the allies the crucial Battle of the Atlantic. Nevertheless, to allow Harris to pound Germany into submission, the Royal Navy was forced to undergo cut backs in air support and the overstretched army's desperate requests for air support were refused.

Wing Commander Guy Gibson's elite 617 Squadron's attack on the Ruhr dams is perhaps the most famous raid of the war, - the "Dambusters Raid".

In terms of technical ingenuity, and courage, this raid by nineteen Lancaster's is without doubt one of the great epic feats of precision bombing and it was Scientist Barnes Wallis who made it possible by developing the *bouncing bomb*. When released, this bomb skipped over several torpedo nets to explode against and breach the dam wall. However, although the damage was spectacular, it was only short-term. The dams were repaired within a few months and water and power supplies were only marginally affected.

Neither the special bomb nor the resulting floods were of any great importance. But, the cost to 617 Squadron was crippling. Nine aircraft never returned and fifty-six highly skilled brave young men were killed. Public Record files now reveal that Harris had received an intelligence report before the raid and knew all about the probability of these losses. The report stated, *"That the dams were not considered a prime objective as the effects were unlikely to be substantial"*.

About one issue, however, there can be no uncertainty. The aircrews of 617 Squadron deserve to be remembered for executing a skillful operation under the most difficult conditions. But, considering the end result and the losses incurred, the raid was deemed a failure and merely bolstered British morale. If that had been the main objective then the raid was an unqualified propaganda success. The very fact that it commands the attention it still does today is proof of that.

By 1943 radar development was advancing rapidly, with each side producing countermeasures to the other. The British "window" innovation for example was incredibly simple; and produced quite dramatic results. "Window" was the name given to the dropping of aluminum tinfoil strips to jam German radar and was used during the attack on Hamburg. This raid was code-named, operation *Gomorrah*, and the choice of this name is particularly gruesome considering that the bomb aiming point was the Church of St. Nicholas.

Over eight thousand tons of bombs were dropped, half of them incendiaries causing fires to merge with existing fires that unleashed a firestorm. Winds of up to 240 km/h. raged through the city and temperatures reached 1 000 degrees C.

The results were horrific; in the intense heat, bomb shelters became vast crematoriums. Fifty thousand citizens perished in the Hamburg devastation; many simply vaporised. A greater tonnage of bombs was dropped and more civilians killed in this single attack than the Luftwaffe managed to inflict on Britain during the entire war. Almost ten square miles of the city lay in ruins and a million panic stricken refugees fled the city. Quite astonishingly, Hamburg recovered with remarkable speed. Within six weeks production output reached the pre-raid figures.

After Hamburg the German air defences underwent a sweeping reorganisation and was now capable of protecting German air space. A fact that Harris chose to ignore because he was determined to flatten Germany and announced that he now intended to wipe Berlin off the face of the earth, which would end the war for the cost of about four hundred RAF bombers. In fact the "Air Battle of Berlin", fought from November 1943 to March 1944 was a disastrous operational defeat and Bomber Command lost one third of its total four- engine aircraft. Enemy fighters came out to intercept them over the North Sea, thus exposing the bomber formations to attack almost the whole way to Berlin and back. This is what faced the brave aircrews in the worst flying conditions imaginable.

No other bombing campaign against a single target was pressed so hard for so long and at such a cost as the attempt to destroy the Reich's capital. When it became clear from reconnaissance that the raids were not achieving much, Harris was ordered by Sir Charles Portal, his commanding officer, to terminate the bombing assault. Records now reveal that Portal gave serious consideration to sacking Harris but flinched from doing so as Harris had direct access to Churchill, visiting him on a weekly basis at Chequers.

Harris had his own agenda, preferring to operate in isolation and in total disregard of Allied strategy. He ignored Portal and in March 1944 launched an attack on Nuremberg. It proved to be an unmitigated disaster with Bomber Command suffering its heaviest losses to date. Enemy night fighters ripped them apart with specially modified twin cannon firing diagonally upwards into the undercarriage of the aircraft. The RAF crews watched in horror, as bomber after bomber went down in flames. More airmen were lost in this single raid than Fighter Command lost during the Battle of Britain. The Luftwaffe decimated Bomber Command, and the only thing that the raid on Nuremberg achieved was to raise the morale of the Third Reich population.

It was one of the great aerial slaughters of all time and signalled the end of Harris's personal attempt to defeat Germany by razing her cities to ashes. Instead of shortening the war by area bombing, Harris had instead crippled Bomber Command. Over a period of just six months eight thousand crewmen were lost, over one thousand aircraft completely destroyed, and an additional one thousand six hundred heavily damaged. These unsustainable losses were vastly beyond British industrial capacity.

But worst of all, every squadron had lost the majority of their highly trained experienced aircrews. Aircrews that were the finest in the world and all of them volunteers, many coming from all parts of the empire including South Africa and Rhodesia. The opponents of Harris's methods, and there now were many, believed (correctly as it turned out) that destroying cities and killing civilians do not win wars, it only serves to harden their resolve. German citizens displayed the same defiance under fire, as the British had during the blitz.

Nothing, however, could sway Harris from his conviction that blasting German cities to kingdom come was the only way to win the war. He was still hell bent on destroying German population areas, even after it had become obvious that this had little effect on their war effort. The harsh facts were that in early 1944 Berlin was virtually unaffected, Hamburg, though severely damaged, had made a remarkable recovery, and the Ruhr's production was accelerating faster than ever. Bomber Command aircrews, therefore, had little to show for their endurance and the loss of so many of their comrades. The German military, industrial, and economic system plus the morale of the people had not been undermined. Harris knew this from intelligence reports, but continued with his area bombing campaign, even though it had degenerated into the random killing of civilians.

The American approach to the bombing campaign was somewhat different. When they entered the war, American strategists considered Britain's aerial bombardment as a complete waste of resources and viewed this as more of a lust for vengeance than a practical military plan of action. They firmly believed that only daylight precision bombing could successfully dislocate the enemy's war effort. By flying high over enemy territory and accurately bombing vital centres, such as factories, power sources, and transportation, not towns and cities, they would successfully achieve this goal with their "self-defending bombers". This is what they called their B-17 Flying Fortress and B-24 Liberator whose massed firepower would provide collective protection.

Aided by the Norden bombsight they claimed they could drop their bomb loads into a barrel from twenty five thousand feet. This boasting came home to roost when they attacked the ball bearing industry at Schweinfurt. Over three hundred German fighters pressed home fierce attacks and American aircraft defence proved totally inadequate. A fearful execution took place and dozens of American aircraft fell from the skies. Nine hundred and fifty highly trained crewmen were lost. The planes that did manage to reach Schweinfurt to drop their bomb loads were then subjected to a relentless pounding on the long flight back.

Sixty aircraft were shot down, eighteen damaged beyond repair, and over one hundred and twenty severely damaged. From the original force of two hundred and ninety one aircraft, no fewer than one hundred and ninety eight had been destroyed or damaged, more than 68% of the entire force. Schweinfurt was virtually unscathed and was back in full production within a few weeks. Determined to succeed, however, the Americans persisted, and from bases in Libya, they launched a strike on the vital Rumanian oil fields of Ploesti. Another disaster; of the one hundred and seventy seven Liberators that took off, fifty four aircraft and five hundred and thirty two airmen never made it back. In addition, fifty five of the returning bombers were so badly shot up, that they were only fit for scrap; a terrible loss rate of over 60%. Ploesti was only slightly damaged and output soon returned to normal.

These catastrophic losses shook the very foundation of American air strategy. The inability of unescorted heavy bombers to get through without taking excessive losses was finally recognised, and even the most ardent proponents were now convinced that there was no such thing as a self-defending bomber. It was obvious that a continued offensive on this scale would not end the war but instead it would obliterate the American Air Corps. The American strategic bombing campaign had failed and deep strikes into enemy territory were suspended. The Reich gained a respite until the long-range escort fighters; the P.47 Thunderbolt and the P.51 Mustang came into service.

The attack on the cultural city of Dresden in February 1945 is perhaps the most controversial of the war. This historic city famed for producing toothpaste, baby powder, and porcelain, was certainly not a military target, and was overpopulated with refugees.

Nevertheless, almost 4,000 tons of Incendiary and high explosive bombs were dropped to create an inferno far worse than the notorious Hamburg firestorm. Dresden burned for seven days, and 80% of the ancient city was gutted. The total casualties will never be known, but estimates go as high as two hundred and fifty thousand. There were not enough able-bodied people left to bury the dead and the devastation was greater than that suffered by any other city during the Second World War, including Hiroshima and Nagasaki. By any standards, the bombing of civilians and refugees at Dresden was an act of barbarism. Undoubtedly, Dresden was a crime against humanity, committed by an insubordinate commander as part of a final destructive frenzy and served no strategic purpose.

When the full horror became known of what happened to this undefended old city, the civilised world was stunned. Churchill did much to distance himself from involvement in the massacring of thousands of civilians, and left Harris to shoulder the blame. The Americans likewise distanced themselves, even though their own planes had gunned down civilians fleeing the burning city the morning after the raid. Due to the public outcry, particularly by the clergy, Churchill ordered an end to area bombing. He wrote to Harris: "It seems to me that bombing German cities simply for the sake of increasing terror should end. I feel the need for more precise concentration upon strategic objectives such as oil and communications rather than mere acts of terror and wanton destruction".

The Americans, although denying it publicly, were essentially no different from Harris. They frequently conducted area bombing, and the reality was that if you were on the receiving end in Germany, you would be unable to distinguish the difference between British and American bombing. It has been said that the British precision bombed area targets and the Americans area bombed precision targets.

What are the more positive aspects of the bombing campaign? It most certainly did tie down enemy manpower and resources. The Germans stationed one million men to defend themselves against the raids and most of the time they were idle, not knowing where and when they were going to be attacked next. Twenty thousand antaircraft guns were allocated to air defence. These guns would have been a godsend to the German armies on the eastern front. More than three thousand five hundred German fighters were shot down over Germany. What if those aircraft had been waiting in France in June 1944?

These are certainly powerful arguments and undoubtedly the impact of bombing was wide-ranging but, the allocation of thousands of anti-aircraft guns, and flak crews to home defence may be misleading. Firstly, anti-aircraft guns cost a lot less to manufacture than four-engine bombers, and few of them were destroyed in battle unlike more than eight thousand RAF bombers. Secondly, the flak crews were often teenagers and women who required far less training than skilled RAF flight crews. There is no doubt that the bombing offensive punished Germany terribly and destroyed centuries old buildings and culture. However, it was not the destruction of cities that brought Germany to her knees.

This only occurred when Bomber Command was ordered to embark on Strategic Bombing of fuel plants and transportation networks. This vital contribution paved the way for the D-Day landings by depriving the enemy of oil and disrupting his communication systems. Prior to striking German synthetic oil plants, production averaged three hundred and sixteen thousand tons per month. By June1944 this fell to one hundred and seven thousand tons and in September, an all time low of just seventeen thousand tons.

Harris was actually incensed when ordered by Supreme Commander General Eisenhower to transfer Bomber Command to tactical operations in support of the D-day operation as it disrupted his programme of slaughtering civilians. However, Eisenhower was correct; the overall effect on the German fuel supply was decisive and undoubtedly the greatest bombing success of the war. Unlike the area raids on cities, the bombing of oil installations severed the jugular vein of the German war effort. If Harris had concentrated on this earlier, it would have grounded the Luftwaffe, thereby; saving RAF crews from heavy losses and in all probability prevented the German counteroffensive in the Ardennes.

Many Germans, including Albert Speer, later wrote that the one thing that could have ended the war in 1944 and saved hundreds of thousands of lives would have been the redirection of Allied bombing from populated areas to oil supplies. Harris's failure to implement the order to attack oil supplies sooner was a major catastrophe and he should have been sacked. If he had been less obsessed with area bombing and if Portal had been stronger in dealing with him, the war would have been shortened by several months. But, all Harris wanted was to bludgeon and pulverise every building in Germany and either kill or leave millions of civilians crippled and homeless. This was his sole priority and he ignored orders to be more flexible, even after command of the air had been won.

He only grudgingly got involved in strategic operations. Yet after the war he claimed the credit for the low D-Day casualties and the destruction of the V weapon sites. It must be said that Harris was not alone when it came to indiscriminate bombing. The Germans also displayed a lack of scruples with their V weapons. Nine thousand V-1's were launched against Britain and these landed anywhere within a thirty mile radius, killing ten thousand people and seriously injuring twenty five thousand.

The last V-2 landed only a few weeks before the German surrender and considering the hopelessness of their position, this was vengeful indiscriminate killing of people, and had no strategic value. In Britain doubts were growing as to the moral justification and indeed even the effectiveness of Bomber Commands area bombing. Immediately after the war the indiscriminate warfare against civilians was questioned and came under severe scrutiny.

Particularly when Harris, despite orders to the contrary, had authorised the devastation of the remaining German cities. He declared, when Germany lay prostrate in the last few months of the war that "Bombing anything in Germany is better than bombing nothing".

This is why the bombing of Dresden in the dying days of the war aroused so much revulsion. Even the troops of the British army of occupation in Germany were dismayed by the total devastation of Germany. Their reports plus pictures and descriptions in the press about what had taken place in Dresden raised questions in Parliament, "Was Allied terror bombing simply a grander version of what the Germans had done to Coventry and Rotterdam?"

If this is so does that mean that Harris was only better at it than the Nazis?

If that is the case, then who benefited from the destruction of Dresden?

The answer to this is that it was the Soviets, and they made full use of it after the war as an example of western savagery. They made sure that the people of their satellites in Eastern Europe were fully aware that the Soviet Union had not indulged in area bombing and had managed to take large towns and cities without blasting them to pieces. They propagated that Soviet forces had simply bypassed populated areas devoid of any military value, and highlighted that the west was no different from Nazi Germany.

To the end of his life in 1984, Harris continued to argue that had he been permitted to continue his assault on the German population in 1944, the war would have ended much sooner. However, incontrovertible evidence proves that the destruction of cities and the slaying of civilians had not reduced essential German war production, nor had it prevented the German armed forces from continuing their operations in Africa, Russia, Italy and Western Europe.

Harris's failure to bring Germany to her knees, and the cost of this failure was evident to all but himself. He was beyond the control of his superior officer, the weak Sir Charles Portal and he frequently defied orders until the very end for the bombers to be directed against specific military and oil supplies in order to continue remorselessly with his personal programme of leveling Germany's cities. If Harris had shown flexibility in the autumn of 1944 and acknowledged that saturation bombing was outmoded and Bomber Command was now capable of better and more important operations, history might have judged him more kindly.

It has been estimated that one-third of British industry was committed to Bomber Command and it absorbed the cream of wartime high technology. The cost of a Lancaster bomber fitted with all its sophisticated equipment was staggering, and by 1944 the total manpower devoted to the production of heavy bombers was equivalent to equipping the entire British Army. Because of the vast resources consumed by Bomber Command, Britain came perilously close to putting all her eggs in one basket and as such was compelled to purchase vast quantities of war material such as tanks, trucks, landing craft, etc. from the United States. In effect, Bomber Command diverted far more resources from Britain's war effort than it did from Germany's.

Area bombing, for all its destructive and terrifying powers, only achieved limited success and was not an economical means to fight a war. After final victory, the whole undertaking was reviewed and considered futile, wasteful, and immoral. The obliteration of Germany's cities in the spring of 1945, when all possible strategic justification had vanished, is a lasting blot on the Allied conduct on the war.

Harris was guilty of the same error as the First World War generals who had been so awed by the effects of their own barrages that they found it inconceivable that the enemy could have survived them. The Bombing offensive is above all, a tale of courage and it is unfortunate that so many fine young men died because their commander was stalemated into a battle of attrition. Harris had persisted with a misguided pre-war doctrine and never grasped the leading principle of air warfare, that is, the winning of air supremacy.

He was ignored in the victory honours list and his request for a special campaign medal for Bomber Command refused. It was made clear to him that he had no further role to play in the RAF and was accused of needlessly sacrificing highly trained bomber crews in the blind belief that saturation bombing alone would win the war. Disappointed by such reappraisal of his war aims and methods, Harris retired to South Africa, where from 1946 to 1953 he was managing director of the South African Marine Corporation.

To many, the conclusion is that the cost of the bomber offensive in life and resources tragically outstripped the results that it achieved.

Despite all the effort and sacrifices by heroic aircrews, final victory would only be achieved by invasion and land battles.

It is a bitter truth that the results obtained by the bombing offensive bore no relation to the cost. Harris's campaign was an aberration. It began as a false strategy and became a moral obscenity.

~THIRTEEN~
STALINGRAD

Stalingrad is arguably the greatest land battle of all time, when in 1942 the Soviets, trapped the German army. The battle would also be renowned for its devastation and ferocity. This battle would not have been necessary had the Soviet Union been destroyed with the single mighty blow as predicted by Adolf Hitler when he launched operation Barbarossa in June 1941.

However, the major German objective in 1942 was switched from capturing Moscow to securing the strategic raw materials vital to the German war effort; mainly the Baku oil fields in the Caucasus. In July 1942, in spite of the setbacks of the previous year, Hitler still felt victory was near and confident that the Red Army would be defeated.

The strength of the Axis powers at the opening of the 1942 summer offensive amounted to two hundred and fifteen divisions, including forty six satellite divisions; the majority from Italy, Hungary and Romania. On paper, this amounted to thirty five divisions more than were deployed at the commencement of operation Barbarossa. But, Hitler chose to ignore the fact that the additional units were mostly of low strength and under equipped.

He was convinced that the enemy having suffered enormous losses was near to complete exhaustion. Hitler was determined to throttle the Soviet state and conceived operation "Blau" which would culminate in a huge pincer movement closing around Stalingrad;

effectively cutting Russia in half. His forces would then thrust into the Caucasus to gain the precious oil fields. Thus, knocking the Soviet Union out of the war before the USA could intervene.

The strategic rationale for taking Stalingrad was to defend the captured Caucasian oil fields. It was also an important communications and arms production centre. However, Hitler would later become obsessed with the symbolism of capturing the city named after his arch foe, regardless of its strategic value.

After the war, German military commanders would claim that Hitler's grandiose plan was never really feasible and totally beyond the capacity of the Wehrmacht. Perhaps they were correct. But, if Hitler's decisive campaign of the war was to have any chance of success then it would require strict adherence to the original plan and the methodical execution of each phase of the operation. And this was precisely what he did not do!

Hitler repeatedly departed from his original plan, constantly interfering and making ruinous deductions; ordering simultaneous attacks or worse still, diverting whole armies from their main objectives. Operation Blau, actually started well, the Russians had expected a renewed thrust towards Moscow, not ever considering an attack on Stalingrad and the Caucasus. Thus, their feeble defences in the south were ill equipped to offer effective resistance and were easily swept aside.

General Paulus's fated 6th Army made good progress, breaking through to punch a huge hole in the Russian lines, and destroying four Soviet armies. Stalingrad lay ahead almost totally undefended; but, there now followed a series of decisions by Hitler to cast grave doubts on his competency as a commander.

Completely confident that Stalingrad would fall easily, he transferred Fourth Panzer Army south to assist in the Caucasus, only to change his mind two weeks later when this move proved unnecessary. At this critical stage of the campaign he also transferred two crack panzer divisions to France, fearing a possible Allied landing. Then he moved two panzer divisions north to counter what proved to be an imaginary threat. Next, he ordered Manstein's Eleventh Army to the north in preparation for an assault on Leningrad.

These transfers reduced, to a bare minimum, the effectiveness of his striking force for the huge task ahead. Hitler's interference in operations became more and more irrational. He constantly underestimated his enemy's capabilities. Rather than argue with their leader the German generals chose to fall in line with his plans, resulting in fatal delays. This caused the build-up of enormous traffic jams, resulting in major logistical and supply difficulties. It was this, more than any other reason that led to the great catastrophe at Stalingrad. Troops were diverted when the city could so easily have been taken before Russian resistance could stiffen. Stalin, now fully aware of his adversary's intentions mobilised all civilians and issued the famous directive: "Not another step back".

The road to Stalingrad had been long and painful and Hitler was completely unaware of the formidable trap into which the German army was about to fall. The Russians were no longer prepared to yield further territory to the German invader. When the German army reached the outskirts of the city the stage was set for what would become the greatest contest of the war.

By September the Germans were engaged in bitter hand to hand combat, fighting for every street. Casualties were enormous and the German army's greatest asset in Russia, rapid movement and mobility, were effectively neutralized and counted for nothing. The concepts of Blitzkrieg were totally irrelevant as the panzers became bogged down in the narrow streets of Stalingrad. Hitler refused to see the consequences of his own folly; more and more troops and equipment required to take the Caucasus were thrown into the cauldron of Stalingrad, only to be decimated.

This was indeed strategic suicide and a battle of attrition resembling Verdun in World War One developed. The German offensive strength was drained; they were fought to a standstill and unable to bring the battle for Stalingrad to a successful conclusion. The errors multiplied, much to the exasperation of the general staff. The German front in Russia was by this stage over three thousand five hundred km's long; and the very configuration invited disaster. The over-stretching of available forces was far beyond the safety margin.

Hitler assumed that the Soviets had no operational reserves left and dismissed German intelligence reports of the massive troop concentrations being built up across the Volga. The Germans had failed to squeeze out the Russian salient's to the north and south of the city which presented the Russians with jumping off points to launch a counter offensive, and General Paulus, the commander of the German 6th Army, had serious misgivings regarding his army's dangerous position.

He attempted to strengthen his flanks which were poorly protected by under-equipped Romanian divisions, the Achilles heel of Paulus's army. He correctly deduced that the Russians may be preparing an assault on his rapidly deteriorating and vulnerable situation and recommended a withdrawal before it was too late. To Hitler, a strategic withdrawal was totally unacceptable and strict orders were issued that the city was to be taken at all cost. Stalingrad had become a matter of personal prestige which riveted his attention, much to the joy of the Russian general staff.

Hitler indicated to Paulus that if victorious he would be promoted to Field Marshall. The general was expected to take Stalingrad or perish in the attempt. By nightfall the Russian assault had smashed through into the rear causing fearful panic. A general rout developed with the Rumanian's fleeing in all directions. The Russians energetically exploited their success and aimed for Kalach in 6th Army's rear.

The following day the Russian armies south of Stalingrad launched their attack with equally devastating effect, and headed for the same destination, Kalach. Soviet General Zhukov had been building up fresh reserves whose morale was high. They were about to encircle and crush the cream of the German army. By mid-November, more than one million men, fourteen thousand heavy guns, nine hundred tanks and one thousand one hudred aircraft were in position to strike the Germans. The carefully planned preparations for the Russian counter offensive were a feat of outstanding organisation, deception and secrecy. Troops had been moved into position, ferried across the Volga at night, observing total radio silence.

The decision to leave their flanks poorly protected by mediocre satellite troops was to have profound consequences. Within three days the pincers had closed, 6th Army was effectively trapped on the Volga. Tenaciously, the Russians proceeded to secure their perimeter against a possible German counter attack from the west. Simultaneously, strong Soviet attacks were launched toward Rostov with the objective of cutting off the German forces in the Caucasus.

Paulus knew that the best option was an immediate break out before the Russians had time to consolidate and secure their perimeter. He urgently requested freedom of action to extricate his forces from the rapidly deteriorating situation. Hitler sealed 6th Army's fate by refusing permission, instead he ordered Paulus to form a temporary hedgehog defence and await help, promising supplies by air and an immediate relief column. This calamitous decision was based on Göring's assurances that the Luftwaffe could supply the five hundred tons of supplies which were required each day to keep 6th Army fighting. This irresponsible pledge by Göring was logistically impossible. Göring's delusions, bordering on lunacy, as far as the German troops at Stalingrad were concerned, grossly misguided Hitler. The actual daily supply never exceeded ninety one tons and the Luftwaffe were to lose five hundred planes and one thousand air crew in the attempt to supply Stalingrad.

Meanwhile Paulus, being a loyal soldier of Prussian descent, dutifully accepted the disastrous order for his two hundred and seventy eight thousand troops in the Stalingrad pocket to await the promised relief. Hitler, now deeply concerned, summoned von Manstein and entrusted him with the mission of relieving beleaguered Stalingrad. Göring, he was assured, would supply Paulus by air until ground contact could be established.

A new army, named "Army Group Don," was assembled under Manstein's command, with orders to drive straight through to Stalingrad and relieve the trapped 6th Army. On December 12th Manstein launched his relief operation called "Winter Storm" on the understanding that Hitler had ordered Paulus to break out to meet him. Only limited success with under strength units was achieved before the attack ground to a halt. Manstein was forced to abandon the rescue mission when a Russian counter stroke threatened him with encirclement. The relief column actually got to within fifty km's, but Paulus did not break out to meet it; Hitler's approval could not be obtained. Manstein, a brilliant strategist, perceived the obvious. He now realised that the chances of relieving 6th Army intact was hopeless.

In fortress Stalingrad the conditions of the stricken army rapidly deteriorated. The exhausted troops were reduced to scavenging and eating horseflesh. Temperatures had plunged to -30 degrees C, and due to Göring's incompetence, hardly any winter clothing had arrived. The men were suffering from frostbite, dysentery and typhus with little hope of survival. The airlift was a chronic failure. Paulus bitterly complained that the Luftwaffe had left him in the lurch.

The Soviets attacked from their perimeter, and remorselessly pushed the German lines further back, capturing most of the advanced airfields from which 6th Army's grossly inadequate air lift was being conducted. Although the fighting remained intense, the end was in sight for the German army at Stalingrad. On January 9th the Russians, in view of the hopelessness of the German position, called on them to surrender and avoid any further unnecessary loss of life. The terms offered were, under the circumstances, considered reasonable. The Russians promised the encircled Germans the guarantee of a return home when hostilities ceased; the alternative was annihilation! Paulus on Hitler's orders rejected them; he was to fight to the bitter end

The questions arise, why did Paulus knowing that the cause was irretrievably lost continue the struggle against hopeless odds, rather than bring the battle to a swift end to spare suffering? Why did Hitler condemn an entire army to death and destruction? The answers lay in the ludicrous disposition of the German army on the eastern front. A far more devastating fate than the loss of 6th Army loomed. Three huge gaps had been opened to threaten Rostov and the only railway supply line for the entire southern wing.

The only real way to halt further exploitation of the gaps torn through the German defence lines was to maintain Stalingrad's pull on the mass of the Soviet forces. Every day the pocket held out, it tied down huge Russian forces; thus gaining the Germans time to stabilise the eastern front. Cruel necessity demanded this last sacrifice of the doomed troops. Otherwise the war was lost!

The unbelievable errors committed in the conduct of operations by Hitler and Göring had undoubtedly brought imminent danger on the German army. Manstein informed Hitler that due to the Stalingrad disaster the position in the Caucasus had become untenable. The German forces had to be withdrawn immediately from the Caucasus. The position of the Wehrmacht had passed from critical to pitiful. Yet, it had to hold on in Stalingrad; surrender would unleash an even greater catastrophe.

Hitler eventually relented, and gave the necessary order. Manstein, by skilful military manoeuvering, managed to evacuate the endangered German armies which Hitler had recklessly and fatally thrust south into the Caucasus. Total disintegration on the eastern front was thus avoided. Manstein, could not save the trapped 6th army at Stalingrad, but he did rescue the armies in the Caucasus from a similar fate.

The Russians at Stalingrad now moved in for the kill against the weak and starving defenders, unleashing a terrifying artillery barrage to take the last remaining airfield still in German hands. Food, ammunition and medicines soon ran out. The Germans fought with desperation and heroism, but could not stop the Russians from breaking through their defence lines. Casualties on both sides were severe. Paulus pleaded with Hitler for permission to surrender. This was again refused. Paulus was however rewarded for his obedience and promoted to Field Marshall. The gesture was transparent; no German Field Marshall had ever surrendered.

Hitler issued dozens of decorations and promotions to the dying survivors in an effort to bolster their resolve. Paulus, verging on a nervous breakdown, finally bowed to the inevitable; 6th Army had disintegrated into isolated pockets with no unified command. The futility of further resistance was acknowledged, and despite his promotion, he capitulated. When Hitler heard the news he flew into an indescribable rage accusing Paulus of dishonour and cowardice, in preferring captivity to suicide.

Hitler's dream of capturing the Caucasian oil was over. He had clearly stated that possession of the Caucasian oil fields spelt either victory or defeat. He was right! The last shot in the battle was fired on February 2nd. 1943. After four months of horrendous fighting, the battle for Stalingrad was over and 99% of the city was reduced to ruins and rubble. Many German soldiers were beaten to death by enraged civilians.

It was a crushing victory for the Soviets and an enormous blow to German prestige. As a result of Hitler's miscalculations, the pride of the Wehrmacht lay scattered in the snow on the banks of the Volga. Rarely has so much been sacrificed at the whim of one man! The shock was immense! He had no one to blame but himself for the German army's greatest defeat. He had taken a great risk in leaving six hundred km's of the German army's northern flank weakly guarded and he paid the ultimate penalty for this vital flaw. Besides the enormous losses in manpower the Germans lost sixty thousand vehicles, representing the equivalent of six months armament production in the Reich.

Stalingrad was the climax of Hitler's eastern campaign. It highlighted the arrogant folly of launching operation Barbarossa, and his overconfidence. He had gravely under estimated the strength of the Soviet Union who could sustain massive losses without breaking. There can be no doubt that German confidence in the outcome of the war was seriously shaken and the invincibility of German arms doubted after Stalingrad.

Hitler's generals became disillusioned in his capacity as a war leader and German morale declined due to his strategic incompetence. The German nation had been denied victory and had lost an army. The Russians had lost many armies; however, German losses were irreplaceable.

The Soviet victory inspired the Red Army, leading to the subsequent Soviet drive towards Berlin. The battle of Stalingrad was the turning point of the Second World War. The strategic initiative had been conceded. Hitler's adventure not only condemned the elite of the German army to death but ensured that the war could not be won. Stalingrad was an irredeemable disaster.

After Stalingrad and particularly Kursk six months later, Hitler's war strategy fundamentally changed. No longer aiming to win the war, he kept going hoping Churchill's Grand Alliance would collapse. He speculated that his enemies would become exhausted, quarrel amongst themselves and eventually agree to a negotiated peace.

Ultimately, he failed to comprehend the determination of the Allied powers, to rid the world of his Nazi regime and all it stood for.

~FOURTEEN~
MEDITERRANEAN NAVAL STRATEGY 1940 – 1943

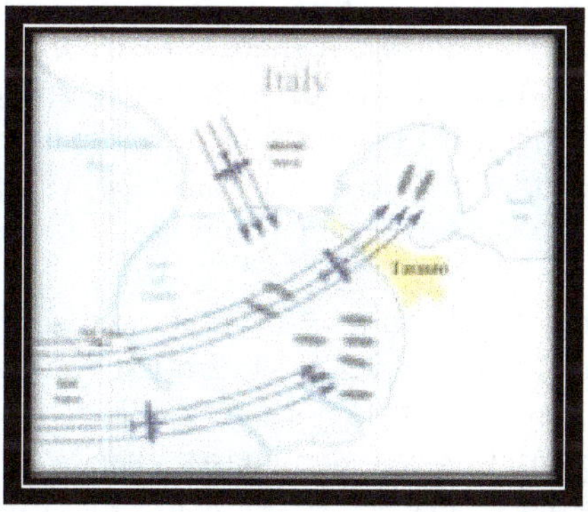

Since the building of the Suez Canal, control of the Mediterranean proved to be vital in maintaining the British Empire's life-line to India and the Far East. Command of the sea ensures that your naval vessels are able to move around at will, while your enemies are forced either to stay in port or to take evasive action.

The British Royal Navy held command of the Mediterranean for over one hundred years, allowing Britain to move troops and supplies in wartime. This came to a sudden end in1940. On 10th June, Mussolini declared war on Britain and France. A week later, the French requested an armistice. The collapse of France meant that French naval control of the western half of the Mediterranean, which had been the agreed Allied strategy, disintegrated.

The numerical balance of naval power in the Mediterranean was altered radically. British strategy now had to be drastically revised. Mussolini referred to the Mediterranean as an Italian lake, and whilst he was desirous of linking Italian North Africa with Italian East Africa. He also had strategic objectives of Egypt, the Middle East oilfields and Malta. To maintain their hold on the region, the British now had to send additional warships to the Mediterranean urgently to protect supplies and reinforcements to Egypt. The modernised battleship HMS Warspite was dispatched as the flagship of Admiral Sir Andrew Cunningham. He was to make his name as the most brilliant Royal Navy admiral since Nelson. Britain's bases in the Middle East depended on the free passage of her convoys between Gibraltar, Malta and Alexandria.

But, Italy was determined to cut them off and maintain her convoy route to North Africa. However, with timely strikes by the Royal Navy and Fleet Air Arm, especially at Taranto, Britain temporarily reversed the situation and shattered Italian hopes of Mediterranean domination.

Admiral Sir James Somerville's 'Force H' was sent to Gibraltar to cover the western Mediterranean. Thus filling the vacuum left by the French. Sadly, Somerville's first duty was that of persuading the French fleet in Algieria to demilitarise itself. When this failed, Churchill ordered Somerville to bombard the French ships. This tragic event took place on July 3rd 1940 and the French suffered the loss of many vessels and sailors. The former allies were embittered, and a very real risk of war between Britain and Vichy France existed. However, it did remove the possibility of these French warships being used against Britain. It also signalled to the neutrals and the United States in particular, that Britain would fight on regardless of the odds.

The first naval clash in this long campaign occurred on 9th July, 1940, when a greatly superior Italian fleet covering a convoy was caught by surprise. The Italians fled but the British were subjected to repeated mass air attacks by Italian High level bombers. Clearly, the Royal Navy needed more air cover. Therefore, the arrival of the armoured-decked aircraft carrier, HMS Illustrious was of most importance.

The next blow against the Italians was to be deadly; an operation which arguably marked a turning point in naval history - the attack on the Italian naval base at Taranto. For the first time in history a battle fleet would be crippled in harbour without a shot being fired by the opposing side's battle fleet. This was carried out on the night of 12th November 1940 by Illustrious, operating twenty one Swordfish biplanes carrying modified torpedoes for use in shallow harbour waters. Codenamed "Operation Judgment", Admiral

Cunningham initially planned to attack on Trafalgar Day, 21stOctober using aircraft from both carriers, the Illustrious and Eagle. Following damage to both ships, the date had to be later changed. But the tenacious Cunningham was undeterred.

While Eagle was still being repaired, he decided to press on using only the Illustrious to launch the air attack against the powerful Italian fleet that included six battleships. The Swordfish attacked in two waves to complete a most successful operation. One battleship was sunk, two battleships and a heavy cruiser, severely damaged. British losses were only two Swordfish. The two damaged battleships were out of action for six months. The sunken battleship *re*quired extensive salvage work and her repairs were still incomplete when Italy surrendered in 1943.

The lesson of Taranto was not lost on the Japanese, who studied the plan very carefully when preparing their air strike against Pearl Harbor a year later. Many other aspects of the raid were also studied; particularly, the modification to the torpedoes. On 27th March 1941, at the impatient urging of their German ally the Italians ordered their fleet to sea to intercept British supply ships running between Egypt and Greece. The Italians were spotted by an RAF Catalina flying-boat. Battleships Warspite, Barham and Valiant, plus the aircraft carrier Formidable put to sea from Alexandria. In what became known as the Battle of Matapan, an Italian Battleship and cruiser were both damaged by torpedo strikes. Two cruisers sent to their rescue were set upon by Cunningham's fleet during the night. The battleship managed to escape, but all three cruisers and two of their accompanying destroyers were sunk. It was a decisive British victory.

Hitler, concerned by Mussolini's lack of progress sent three hundred aircraft to Sicily, including Stukas, specially trained in shipping attack. Enemy air attacks sank the cruiser Southampton and damaged the Gloucester. The armoured flight deck of Illustrious saved her from destruction but several bomb hits made it necessary to send her for repairs.

Another vital strategic consideration was Gibraltar. "The Rock" had been a British fortress since the 18th century and played a crucial role in British naval strategy. It provided a strongly defended harbour from which ships could operate in both the Atlantic and the Mediterranean. Malta was in a different situation, and despite its strategic importance, had been neglected. Air defence comprised of six obsolete Gloster Gladiator biplanes. It was not until sixty one Spitfires were flown in from aircraft carriers that the air defences strengthened.

Although heavy Allied casualties were inflicted and many ships were sunk or severely damaged, Malta survived to receive badly needed supplies. A further eighty two Hurricanes were flown off from the Ark Royal in March, April and May 1941 to prevent Malta's complete neutralisation. As a result, British submarines and destroyers based at Malta could continue to operate and succeeded in intercepting and annihilating enemy convoys.

The fleet was next called upon to cover the passage through the Mediterranean of the 'Tiger' convoy carrying desperately needed tank reinforcements to Wavell's army in Egypt. The Italian's did not even venture out to intervene. As a result, on 12th May, two hundred and thirty eight tanks and forty three crated Hurricanes were safely delivered at Alexandria.

The Mediterranean Fleet was then forced to turn its whole attention to the evacuation of the army from Greece. Though fifty thousand troops were safely embarked, four troopships and two destroyers were sunk by dive-bombers. The fateful decision was then made that Crete, without any air defence was to be held. In Greece, the Germans assembled two air fleets including paratroops to invade Crete. The German attack began on 20thMay, 1941. It was clear from the start that the navy would have to take the brunt and sustain heavy losses. So it proved: the cost to the fleet was three cruisers and six destroyers sunk. Amongst them was the gallant HMS Kelly of Lord Mountbatten fame. The main strategic consequence of the loss of Crete was that the German Air Force was now able to attack shipping passing to and from Egypt.

When Hitler launched the attack on Russia in June 1941, much of the German air force was transferred east. The neutralising of Malta was left to the Italian Air Force. A task they were unable to fulfil, allowing the island's air and sea offensive capability to recover. This resulted in a steady attrition of Italian shipping that had a marked effect on their supplies to North Africa.

With the relaxation of the air assault on Malta, two squadrons of Blenheim bombers could also be deployed on the island. These aircraft bombed at masthead height by daylight and inflicted a fearful casualty rate. The pendulum of fortune had swung back again in favour of Britain whose convoys reached Malta safely and the island was now stocked until the following spring. This success was followed by the arrival at Malta on 21stOctobert, 1941 of two cruisers and two destroyers. During the night November 8/9th they intercepted an Italian convoy of seven merchant ships and, without loss to themselves, destroyed them all. This played a large part in the fuel and supply shortages Rommel had to cope with, and virtually brought a halt to his operations.

Malta's submarines sank three hundred thousand tons of Axis shipping between July and September 1941, a situation described by the Germans as 'catastrophic'. In November 1941, they destroyed 63% of all Axis cargo ships bound for North Africa. However, the pendulum that had swung towards British success, as frequently happens in war, was soon to be reversed. The Royal Navy was to suffer several disastrous setbacks.

Dismayed by the defeats experienced by his Italian ally, Hitler ordered the deployment of ten U-boats to the Mediterranean. They quickly achieved two resounding successes. On November 13th, the Ark Royal, returning from delivering Hurricanes to Malta, was torpedoed and sunk off Gibraltar. Twelve days later, at the other end of the Mediterranean, a U-Boat attacked and blew up the Battleship Barham. Of the five battleships lost by the Royal Navy during World War Two, Barham was the only one to be lost at sea to a U Boat.

Meanwhile, Hitler ordered an air fleet to be transferred from the eastern front to Sicily with a view to finally neutralising Malta. By the end of 1941 the renewed 'Blitz' was well under way. However, there was some success for the Royal Navy. Three British destroyers intercepted, torpedoed and sunk two Italian cruisers off Cape Bon. At the same time, an Italian battleship was torpedoed by a British submarine and put out of action for several months.

While this was occurring, three Italian underwater chariots transported on a submarine penetrated Alexandria harbour to attach their explosive charges to the battleships Queen Elizabeth and Valiant. This brilliantly executed feat resulted in both ships sinking onto the shallow harbour bottom. In a few disastrous days, Cunningham's fleet had been reduced to a few destroyers, three light cruisers, an old anti-aircraft cruiser and the light cruiser Ajax.

British sea power in the Mediterranean reached its nadir, and the time of Malta's greatest suffering was then to come. The Luftwaffe raided Malta almost every day between January and July, neutralising warships and aircraft. Axis supplies flowed again, allowing Rommel to advance to El Alamein.

Damage to the Maltese capital Valetta was severe. Between March and April 1942, Malta received twice the tonnage of bombs dropped on London at that time. One thousand five hundred of her citizens were killed and four thousand wounded. The island's population was later awarded the George Cross in recognition of the heroism of its people.

At this stage, the Axis powers were winning the war at sea in the Mediterranean and on the ground in North Africa. Malta was under constant siege and Rommel's supply ships were crossing the Mediterranean almost unopposed. It took determined efforts, at a great cost in men and ships, to save Malta. Allied control of the Mediterranean was gradually regained, mainly due to advances in technology and intelligence. This was undoubtedly the underlying cause of victory at El Alamein, and the Royal Navy's contribution to the 'Desert Victory'.

Operation Torch took place on November 1942, when Allied ground forces invaded north-west Africa. Following the Axis surrender in North Africa in May 1943, the Allies invaded Sicily in July. The Allies then invaded Italy on 3rd September 1943, and the Italians surrendered on 8th September. The Allies continued to push the Germans northwards through Italy for the rest of the year; thus bringing the Allied Mediterranean strategy to a successful conclusion.

~FIFTEEN~
OPERATION TORCH

Operation Torch, the Anglo- American invasion of north-west Africa took place on 8th November 1942. This operation which still commands incredulity came a fortnight after the launching of the British El Alamein offensive at the opposite end of the Mediterranean. By the time that the new Allied expedition landed in Morocco and Algeria, Rommel's Afrika Korps was in full retreat from Egypt.

Torch had been in preparation for only three months, but in conception much longer. At the Arcadia Conference in Washington, the first Allied meeting following Pearl Harbor, Churchill put forward the 'North-West Africa Project' as the first step towards "closing and tightening the ring around Germany". However, Roosevelt's service advisers were dubious about its practicability preferring an early and more direct attack against Hitler's Europe. They were of the view that this diversion to the Mediterranean would only prolong Allied efforts; preferring an early cross-Channel attack to be launched by August or September 1942. This was aimed at meeting Stalin's demand for the opening of a 'Second Front' to relieve pressure on Russia.

The President and his strategic advisers had already decided on a "Germany first" strategy. The United States would remain on the defensive against Japan while exerting every effort to defeat Germany and of course Italy. They could then turn to the Pacific and eliminate Japan. This commitment to Germany First was heartening to Churchill, but the enthusiasm to fight the Germans immediately was unrealistic. The British Prime Minister emphasised the drawbacks of a premature landing in Europe with inadequate strength; pointing out the risks of being overwhelmed, without bringing any appreciable relief to the Russians.

He was aware that American military forces were in the process of expanding, organising, and training for combat; therefore, they were hardly a match for their strong and veteran foe; as Dieppe was later to prove for the Canadians. Churchill cabled to Roosevelt that the plan for a landing in France in 1942, should be discarded, and went on to urge, once again, the case for Torch as the sole means by which the U.S. could strike at Hitler. The American Chiefs of Staff reacted with renewed objections.

Then, a major event occurred that was to change the entire situation. Rommel's unexpected counterstroke dislocated the 8th Army's westward advance and forced the British to retreat more than two hundred miles to the Gazala Line. This was followed in June by the collapse at Tobruk. So, instead of advancing westward as planned, the 8th Army was thrown back in disorder a further four hundred miles before halting at Alamein. Alamein was the last possible stop-line short of Alexandria, Cairo, and the Nile Delta. Churchill was in Washington with his Chiefs-of-Staff when he learned of the disaster at Tobruk. A joint Allied operation in North Africa was now viewed as essential. The operation was approved and Roosevelt promised to provide three hundred Sherman tanks to the 8th Army.

Considering the long ties of Franco-American friendship Operation "Torch" would have to be in essence an outwardly American show. The French still remained bitter about the British attack on their fleet shortly after France fell and would certainly oppose a British landing. The initial landing waves would therefore consist solely of American troops; also the commander of the overall operation would have to be an American.

For this reason, Eisenhower was appointed the Allied Commander-in-Chief. He had never been in combat, but had impressed his superiors with his thorough grasp of military matters, and his ability to make people of different backgrounds work together. Yet he was relatively unknown and "Torch", a complicated venture to be undertaken in considerable haste, would be a serious challenge. As it turned out, he grew in stature and self-confidence resulting in a meteoritic military and later political career.

Eisenhower chose Major-General Mark Clark as his Deputy Commander-in-Chief. Clark would prove invaluable in dealing with the French in North Africa. Lt. General Kenneth Anderson was to command British ground forces and the flamboyant and charismatic Major-General George Patton the American. Patton was aggressive and experienced in combat, and he would soon gain the reputation as America's best fighting general. A Combined Planning Staff, responsible to Eisenhower was established with the overall strategic objective of gaining control of North Africa from the Atlantic to the Red Sea.

Torch had profound political implications that were to influence the campaign. The Americans had hopes that the French would greet them as liberators. But, in the event that they resisted, Allied forces would have to overwhelm and defeat them. This was one of the great imponderables of Operation Torch.

The next stage was to co-ordinate plans for Eisenhower's forces to push into Tunisia and link up with the British 8th Army under Montgomery. This would entrap and eliminate Axis forces, thus giving the Allies complete control over the northern shore of Africa. It would open the possibility of further operations across the Mediterranean into the European continent and secure Middle East oil supplies for the Allies.

Exactly where to land had to be measured against the considerable threats posed by U-Boats in the Atlantic and Mediterranean, plus enemy aircraft operating from Sicily. The close proximity of Tunisia to Sicily made it extremely likely that German and Italian forces would be dispatched to counter the Allied landings. To limit such action, the major part of the invasion force would have to take place in the Mediterranean. The danger here was that the Strait of Gibraltar might be blocked to Allied shipping. Either by Spain or Hitler's forces driving through Spain.

Close coordination among Allied land, sea, and air forces would be essential. Major air units included one fleet carrier and four escort carriers, which embarked one hundred and eighty fighters, thirty six dive bombers, and twenty six torpedo bombers. The entire task force also included three battleships, seven cruisers, thirty eight destroyers, four submarines, thirty troop transports, plus numerous support vessels; a vast undertaking of over one hundred thousand men.

"Torch" would consist of three major landings; the Western, Centre and Eastern Task Forces. The Western Task Force was to be wholly American and would sail across the Atlantic from Norfolk, Virginia. Patton would sail on board the flagship, USS Augusta, and planned to come ashore near Casablanca in French Morocco.

The Centre Task Force would consist of American troops transported from the United Kingdom to Oran in Algeria. The Eastern Task Force was also formed in the United Kingdom and would be predominantly British. Making the initial landings near Algiers, however, would be a relatively small American force. Air support during the first few critical days would be from the carriers. Once the French airfields came into Allied hands, carrier borne aircraft would be flown to those airfields.

Regarding the complicated political situation; officially, French leaders in North Africa pledged their support for Vichy to defend their colonies against any attacker. But, covertly, many of them conspired against the Axis; realising that the only chance of liberating their country was through an Allied victory.

Robert Murphy, the chief American diplomat in North Africa, had been actively engaged in discreet meetings with French officers sympathetic to the Allies. As a result, the French now urged that a senior Allied military representative should come secretly to Algiers for talks. Accordingly General Mark Clark flew to Gibraltar with four key staff officers. The party was then transported by a British submarine to a rendezvous west of Algiers. Clark told the French that a large Anglo-American force was being dispatched to North Africa. In the interests of security, however, he abstained from giving details of the time and places of the Allied landings.

An important issue discussed was the choice of the most suitable French leader to rally the French forces in North Africa to the Allied side. Admiral Darlan, Vichy's second in command, had by chance happened to be in Algiers visiting his sick son in hospital. Darlan was the highest governmental official on the scene, and he represented the legal authority of France. He was, however, anti-British due to their previous action against the French fleet. This rendered him doubtful in view of the fact that the British were playing a major role in Torch.

De Gaulle was also ruled out. Roosevelt had developed a deep distrust of him and disliked his arrogance. Roosevelt actually insisted that all information about Torch should be withheld from de Gaulle. In these circumstances the Americans, from the President downward, readily accepted the view that General Giraud was the most acceptable candidate for the leadership of the French in North Africa. Giraud, had been taken prisoner in May 1940, but had managed to escape and reach North Africa. Here he met many officers, who shared his desire to liberate France.

As previously mentioned, the Allies hoped that the French would not resist. For that reason there would be no pre-assault bombing or naval bombardment. A system was, however, devised to allow any unit to announce that it was being fired on by announcing "batter up," which meant it was preparing to return fire in self-defence. Only the task force commanders could initiate the general engagement of French forces. That command was "play ball."

The western task force route took it south of the Azores and had the good fortune that no Axis submarines spotted it. By November 6th as the task force approached the Moroccan coast, the weather took a turn for the worse. It was nevertheless decided to keep to schedule, despite a heavy surf along the coast. The Western Task Force split into three attack groups and took up positions off the shore line defended by coastal defence guns, and the 15-inch guns of the battleship Jean Bart.

Once the beach heads were secured, they would break out and advance on Casablanca. A major threat was the French fleet based less than fifteen miles from the landing beaches. It consisted of a light cruiser, ten destroyers, and eleven submarines. The U.S. Navy, therefore, placed its most powerful ships close to the Assault Groups. This included the new battleship Massachusetts armed with nine 16-inch guns. Heavy cruisers armed with 8-inch guns, including Augusta with Patton aboard, plus light cruisers and destroyers.

As the first waves hit the beaches, shipboard loudspeakers blared across the water in French "Don't shoot, we are your friends, we are American". But the French did resist the landings, and the sea erupted from coastal defence guns plus Jean Bart firing on U.S. Ships. The French Fleet then attacked and several U.S. vessels were hit. At 04h 38, the attack Group Commander signalled "play ball" and the US ships immediately engaged the French. In the first minutes of the exchange the Massachusetts shelled the main coastal defence battery, effectively silencing it.

The French came off worst in the engagement. By early afternoon five French destroyers had been sunk, and a cruiser driven ashore. Additionally, Jean Bart's main guns were put out of action following several hits from the Massachusetts. More troops stormed ashore north of the harbour and began to fan out while simultaneously overrunning artillery positions. Overhead there were numerous dogfights between French and U.S. aircraft. Dive bombers sank three French submarines in the port. The threat of a French naval attack was eliminated.

On the landing beaches many mishaps and delays arose due to inexperience. Although the advance on Casablanca got going on the second day, and met no serious opposition, it was halted due to lack of equipment, which was piling up on the beaches. Even on the third day little progress was made. And it was not until evening that the airport was captured.

With over five thousand French troops in or near the city, and reinforcements on the way, there was the potential for a major battle near the heavily populated city. But by the morning of November 10th, French opposition began to collapse. Clark pressed Darlan to order an immediate cease-fire everywhere in French North Africa. When Darlan hesitated, Clark pounded the table and declared that he would get Giraud to issue the order.

Darlan pointed out Giraud's lack of legal authority, and insisted that such an order would result in the immediate occupation of southern France by the Germans - a forecast that was soon borne out. Clark then told Darlan that unless he issued the order immediately he would be taken into custody. Darlan after a brief discussion with his staff accepted this ultimatum and issued the cease-fire order. The main foe now became German U-boats. On the evening of 11thNovember, several U-Boats slipped among the transports and sank four that were still loaded with over 90% of their supplies.

Meanwhile, the Americans were landing at Oran. The plan was to capture the port by a double envelopment. The operation started well, surprise was complete and no opposition was met on the beaches. The advance from the beach-heads got going and soon reached the airfield that was readied to receive aircraft. By 10h 30 the first carrier borne P–40's landed at the airfield and the Port was firmly in American hands. American casualties totaled two hundred and seventy five killed, three hundred and twenty five wounded, and fifteen missing.

Algiers was the most important objective of "Torch" as it was closest to Tunis. Following a small American detachment, the first wave of British troops came ashore west of Algiers. The operations went surprisingly smoothly. French units in the area reported they had been ordered not to resist and on 8th November, the British had taken their objectives.

By this stage Hitler ordered his forces to invade Vichy France, and capture the French Fleet.

The Axis powers were now, as anticipated, dispatching forces to Tunisia. German and Italian aircraft started to arrive near Tunis and heavy equipment was brought over to Bizerta. For a hastily improvised move, at a time when Axis forces were hard pressed everywhere, this was a notable achievement. The German quick reaction was extraordinarily effective, moving troops into the unoccupied zone of France and simultaneously strong forces were flown to Bizerta. How did Rommel feel about this after being denied reinforcements for so long?

Darlan meanwhile had made a detailed agreement with Clark for cooperative action. Then, on Christmas Eve, he was assassinated by a young man believed to be a fanatical Gaullist. It must be said that if Clark had not succeeded in enlisting Darlan's help, the operation would have been much tougher than it turned out. There were almost one hundred and twenty thousand French troops in North Africa that could have provided formidable opposition. Another cause of relief during this critical period, and especially the first few days, was that Spain had abstained from any intervention and Hitler had not attempted to strike through Spain against Gibraltar. Another important factor is that, en route to French North Africa, the Allies heard the news of the victory at EI Alamein.

The success of the 8th Army immediately placed expectations that they would unite with Montgomery's desert veterans to destroy the Germans and Italians before Christmas. Unfortunately, these hopes were proved unfounded. Northern Tunisia's, airfields and ports were readily accessible to the Axis. Tunis was five hundred and sixty miles away from Algiers over mountainous country across which there were only two metalled roads and an indifferent railway.

The German reaction to the landings was characteristically rapid. Full credit must go to Field-Marshal Kesselring, commander of German forces in the Mediterranean. Though Hitler and his high command remained unperturbed by the growing weight of intelligence reports that an Allied landing was imminent; Kesselring was more alert. In October, he had asked for troops to be concentrated in southern Italy as an immediate reserve to counter any landings made in the Mediterranean.

Again in early November, he made requests, but was refused by Hitler who considered that Kesselring's existing forces were sufficient to repel any landings. Denied assistance, Kesselring brought to readiness his troops and concentrated a mixed number of air and sea transports. Later, he was able to call on the biggest air transport of all the Me-323. But this was only when the landing forced Hitler, and the German High Command to support him.

The news of the landings reached Kesselring within hours on November 8th. And even before Hitler gave him a free hand to act in response, Kesselring had set in train immediate countermeasures. The first German soldiers were deployed covering the approaches to Tunisian airfields, and bombers were already attacking Allied targets in Algiers.

Given the distances, the poor roads, and the rough terrain, the Eastern Task Force, predominantly British, initially made excellent progress. By 25th November they were in contact with Axis units and attacked. But they were at a considerable disadvantage. Their lines of communications were weak and air support irregular. Eisenhower sent reinforcements, but on 24thDecember, after visiting the front, he had to agree that an immediate attempt to capture Bizerta and Tunis would have to be abandoned. They had outrun their supply lines, and lacked the impetus to break through.

A stalemate now set in to bring the initial stage of Torch to an end. The Germans aided by the weather and the terrain, were able to stabilize a line in the mountains west of the Tunisian Plains; thus frustrating Allied hopes of capturing Tunis and Bizerta by Christmas. The stalemate was disappointing. But much had gone well, the intricate organization had been effective and had proved it possible to form integrated American and British staffs.

The landings came as a complete surprise; mainly due to effective intelligence and the fact that Italian surface ships were unable to leave port because of lack of fuel. Spain had not interfered, and French troops, re-armed with American equipment, were now Allies.

One of the weaknesses of the Torch planning was that no provision had been made to neutralise Tunisia during the initial stages. Perhaps a little more boldness in the original concept of the operation may possibly have allowed the capture of Tunisia. German competent reaction was, however, faster than Allied planners had thought possible, resulting in a race developing to seize vital airfields and occupy the passes leading into the Tunisian plain. Meanwhile Rommel had been conducting a well organised fighting retreat from Alamein.

Despite the aspirations of the US Chiefs of Staff, Torch did not initially bring American troops into contact with the armed forces of

Germany. That confrontation would only take place in 1943. But the quick and perhaps injudicious success over the French resulted in American overconfidence, even arrogance.

Having observed this, Rommel and his commanders were openly contemptuous of the Americans' ability to fight. After a buildup that included heavy Tiger tanks mounting the 88mm gun, the Afrika Korps shattered the Americans at Kasserine Pass and drove them back. Kasserine Pass was a bitter introduction for the Americans and would teach them much about the Wehrmacht.

The American M3 Lee and Grant tanks had a high silhouette and were difficult to operate in combat. They were no match for the German Mark IV and Tiger Panzers. American tactical doctrine was inflexible and did not account for the rapid German advance. As a result, the Americans suffered heavy losses at Kasserine, one thousand, hundreds taken prisoner, and the loss of most of their heavy equipment.

The Germans analysed captured American equipment and sent back unfavorable reports, which would entice German commanders to underestimate them in the future. Whereas, the Americans studied the action of Kasserine Pass more intently than the Germans. This resulted in a change of leadership and gave junior officers the authority to make on-the-spot decisions.

Major General Fredendall was replaced by the more aggressive Patton. The M3 tank was quickly replaced with the M4 Sherman. While it was never the equal of the German tanks, it was easier to maintain and had greater range. Most importantly Kasserine Pass taught the Americans the doctrine of massed firepower, and to coordinate aircraft with ground forces.

Montgomery's Eighth Army attacked from the east and in a series of probing battles weakened the Axis forces. There was a break-through at the Mareth line and the Allies linked up on April 8th. Rommel, a sick and exhausted man, had already been flown out, too ill to continue the battle. The besieged Panzerarmee was now confronted by the Royal Navy that was gathering for "Operation Retribution." A Dunkirk in reverse for which Admiral Cunningham had issued instructions – Sink, Burn, Destroy, let nothing pass. The Axis forces were rapidly running out of food, ammunition and fuel. By 13thMay, the day the last resistance ended, two hundred and forty thousand Italian and German prisoners were rounded up.

Years of hard fighting lay ahead, nevertheless, Churchill was indeed accurate after Alamein when he predicted this was "not the beginning of the end, but the end of the beginning".

During the entire North African campaign, the Germans and Italians suffered six hundred and twenty thousand casualties, while the British Commonwealth lost two hundred and twenty thousand men. American casualties in Tunisia alone totaled more than eighteen thousand five hundred.

The Allied victory in North Africa opened a second front against the Axis to remove the threat to British supply lines to Asia and Africa. It was critically important to the course of the war. The Torch landings represented the first major British-American combined offensive, and it set the pattern for Allied unity and cohesion in subsequent coalition ventures.

Largely improvised, "Torch" was a triumph of planning and execution. It required an unprecedented effort to build up an American assault force in the United States, separated by three thousand miles from the other two assault forces; then to arrange for the entire task force to converge simultaneously on the North African coast. Torch was the largest joint amphibious operation undertaken up to that time. Thus it was in many ways a watershed event.

The number of issues that had to be considered and resolved was enormous. The joint services had never conducted an operation like this. The coordination and control of naval gunfire and air support was also in its infancy during Torch. The capabilities of land, sea, and air forces had to be carefully orchestrated, especially since the enemy was, at least on paper, numerically equal or in some areas superior.

It was the overall capability of the joint force that resulted in the timely accomplishment of the mission. Certainly the inexperienced forces that took part in Torch were fortunate that the French did not put up protracted resistance. Had the Vichy troops been more determined, had the Navy not been able to quickly suppress their defences, or had the enemy been better alerted, the landings could have been much more costly. The United States could not afford a defeat at this stage of the war. Strategically, it would have been an enormous setback if their first offensive had ended in defeat.

Torch still offers many examples of what a joint force must do to make an operation successful. Probably the most difficult mission that the Armed Forces will be called upon to perform in the future is a long distance operation against a competent opponent. That is exactly what happened in Torch.

Finally, "Torch" was the first of a series of large-scale coalition amphibious landings, Sicily, southern Italy, southern France and Normandy would lead the Allies to the final battle with the enemy.

~SIXTEEN~
THE RESCUE OF MUSSOLINI

It was frequently stated by Adolf Hitler that even Mussolini could never make anything but Italians out of the Italians. Goebbels, the Nazi propaganda minister, remarked that Mussolini did not deserve the Italians and the Italians did not deserve Mussolini. That is what his allies had to say, so what about his opponents. Sir Winston Churchill for example once stated that if he were an Italian he would readily have donned the black shirt of Il Duce's fascist party. Perhaps even more significantly Mahatma Gandhi claimed that Benito Mussolini was the greatest leader in the world. Therefore, would it appear that Mussolini's influence and status has often been understated or perhaps not given any credit at all?

So who was this enigmatic 20th century dictator and what was his role, particularly in the Second World War? In an attempt to come to terms with this it should be understood that from the beginning Mussolini, considered himself to be a self-styled modern day Caesar who in 1940 became dazzled with the prospect of quick and easy victories. As a result he took a totally ill prepared Italy to war with the objective of emulating Hitler's military triumphs; mistakenly believing that Britain, like France was doomed to defeat.

He expected swift military conquests in Egypt, Greece and East Africa. However, a string of disastrous campaigns necessitated German military intervention, serving to confirm Italy's position as the junior axis partner. Furthermore, the failure of Mussolini's military campaigns made the war unpopular in Italy, and this was exacerbated when he sent Italian troops to the eastern front and when he declared war on America. However, he chose to ignore the warning signs, because Mussolini's ego was greater than his judgment.

This was most apparent in June 1942 when Rommel appeared to be winning the battle for North Africa. Mussolini accompanied with a white stallion promptly flew to Tripoli intending to lead the victorious Italian forces into Cairo. However, Rommel's defeat at El Alamein and subsequent allied landings in French North Africa, coinciding with the German disaster at Stalingrad forced the dictators on to the defensive. The reverses they had suffered effectively destroyed their aim of bringing the war to a victorious conclusion by 1943.

Italy had floundered everywhere, and the Germans were stalled on the Russian front. The Italian people had become fiercely anti-German and transferred this hatred onto Mussolini. As a result, many influential Italians wanted an end to the fighting, and this was intensified when allied bombing and tight food rationing added to the war weariness of the population. Two hundred and fifty thousand Italian casualties had been sustained in Russia and a further two hundred and fifty thousand in Africa. On top of this, 75% of the Italian merchant fleet was at the bottom of the sea. This was at a time when American and Soviet overwhelming resources were being brought to bear on the Axis.

It was only Hitler, and a few of his henchmen in Berlin who believed that they could win a war on two fronts. Hitler attempted to justify this when in February 1943 he explained to an unwell Mussolini that the Russians had to-date suffered eleven and a half million casualties while the Germans had lost only one and a half million.

Hitler conceded to Mussolini that the British and Americans had achieved temporary advantages. However, his U Boats would rectify this by destroying their supply lines. But, despite these assurances by Hitler, Mussolini was not persuaded. He was convinced that the allies intended to invade Italy. The grand illusion of a short and victorious war was effectively over, and Mussolini pleaded with Hitler to transfer Axis troops from Russia to North Africa with the object of inflicting a major defeat on the allied forces.

Mussolini urged Hitler to come to an agreement with Russia, otherwise withdraw the Axis front line troops to pre-prepared eastern front defence lines that could be held with smaller forces. The Italian dictator had by this stage become only a shadow of his former self. He now realised that not only had the myth of German invincibility been shattered but also there was no hope that he could separate Italy from the dictates of the Third Reich.

After the allies landed in Sicily in July 1943, the king of Italy decided that if Italy was to be spared from further ruin then Mussolini would have to be removed from power. The populace had completely turned against him and demonstrated this with industrial strikes against the fascist regime. On July 26th Mussolini was dismissed by King Victor Emmanuel and for security reasons taken by ambulance to the military police barracks. Marshall Badoglio was declared head of state. Mussolini's removal caused a sensation throughout Italy, and fascist symbols and effigies of Mussolini were either destroyed, or defaced.

The Germans were initially taken by surprise in Italy. But soon recovered and reacted with typical efficiency and promptness and Hitler immediately plotted Mussolini's rescue. Meanwhile, the new Italian government fearing German retribution issued a proclamation that Italy would continue on the side of Germany. The Italians, true to form, changed sides, and on 3rd September 1943 Italy unconditionally surrendered to the Allies, informing them that Mussolini would be handed over to be tried for war crimes.

They requested that the American 82nd airborne division be parachuted into Rome to bolster their defence and prevent a German takeover. But the Germans moved too quickly and the drop was cancelled. The Italian government was forced to flee to the Allied occupied south and Italy was plunged into tragedy, finding itself in the invidious position of being occupied by opposing armies.

The six-week delay that elapsed between the fall of Mussolini and the announcement of the surrender had allowed the Germans plenty of time to reinforce their troops in Italy. This was demonstrated in the north where Rommel caught the Italians completely off balance and disarmed ten divisions. Many Italian soldiers simply discarded their uniforms, hid their weapons and just melted away. But six hundred thousand Italians stationed in the Balkans and Crete were given the stark choice of either fighting alongside the Germans or working as forced labour in the fatherland. Few chose to fight.

Meanwhile Mussolini's Italian captors moved him half a dozen times to foil and frustrate his German rescuers; but, security was lax and an intercepted radio message soon put the Germans on his trail. He was traced to a remote hotel on Gran Sasso, a mountain top ski resort deep in the Apennines that could only be reached by cable car.

Captain Otto Skorzeny of the recently formed SS commando group was summoned by Hitler and ordered to rescue Hitler's longtime friend and fascist comrade. Skorzeny was later to be known as the most dangerous man in Europe and had three alternatives to evaluate; a ground attack, a parachute drop, or a glider assault. The ground attack was ruled out owing to the number of troops required. A parachute drop was impractical due to the thin air at that altitude, which would cause paratroopers to descend too fast and plunge to their deaths. Therefore, the glider assault was considered as the only feasible solution.

On 12th September German troops seized the cable car station at the foot of the mountain while ninety glider borne commandos led by Otto Skorzeny crash landed within fifteen metres of the hotel. Skorzeny had placed an Italian general in front of his men shouting an order not to open fire. This confused the defenders and the hotel was successfully stormed. The guard commander surrendered and ordered a white bedspread hung from the hotel window to signify this to the remaining Italians on the mountain top. Mussolini had been rescued unopposed. Skorzeny approached him and announced, "Duce, the Führer has sent me, you are now free".

Skorzeny considered it would be too risky to try and get the now haggard looking Mussolini through the Italian lines by road. Therefore, the six foot four Skorzeny decided to fly off the mountain with Mussolini in a light Fiessler Storch aircraft. The heavily overloaded plane lurched off the edge of the mountain, and plunged down into the valley before the pilot could gain control and fly the aircraft to a Luftwaffe airfield. Mussolini was then flown to Vienna and finally to Hitler's headquarters at Rastenburg. This exploit earned Skorzeny worldwide fame; he was feted by the Nazi hierarchy, promoted to Major and awarded the Knight's Cross of the Iron Cross.

Mussolini, with Hitler's backing decided to establish a new fascist republic in German occupied northern Italy. He was installed as the puppet ruler of this tiny state that was only recognised by Germany and Japan until the end of the war.

On his return to power, Mussolini set in motion the arrest of most of those responsible for the coup that had ousted him from power. One of those found guilty and subsequently executed was Count Ciano, his so- in- law and former foreign minister.

In April 1945, with the Axis armies collapsing and Allied victory imminent, Mussolini received information that the Germans in Italy were discussing surrender terms. Angered and disillusioned by what he perceived to be betrayal and rejecting advice to flee to Spain, he decided that he and his mistress Clara Petacci, who had recently joined him in Milan, should attempt to escape to the Alps in a German convoy.

Within a few days, however, communist partisans halted the convoy at a road block. They would only allow Germans through but not Italians, and fascists in particular. Mussolini had tried to disguise himself by donning a German uniform and hiding among the German troops. But during the search the partisans recognised Mussolini and his mistress. They were both arrested and after a brief incarceration, shot down at the road side.

The bodies of Mussolini, Clara Petacci and other fascist leaders were then taken back to Milan and dumped in front of the same garage that Mussolini the previous year had ordered twenty partisans executed. Irate crowds gathered around the corpses and mutilated them. After lying in the street for several hours the bodies were strung up by their ankles for the edification of the crowd. In a grisly display people hurled filth and spat at the disfigured corpses. One vindictive woman fired five shots into Mussolini to avenge her five dead sons.

When Adolf Hitler heard of how Mussolini had met his end and been put on public display, he vowed this would not happen to him. Hitler and his new wife, Eva Braun committed suicide. Their bodies were placed in a shell hole outside of his Berlin bunker and cremated.

Returning to Mussolini, his life was ruled by the drive for personal glory. His aggression plus constant indecision ultimately destroyed both him and his regime. If he had not felt the need to gain more territory and compete with Hitler he probably would have continued to stay in power.

The other main player in this drama was Otto "Scarface" Skorzeny fought with distinction in campaigns in the west and on the eastern front. He was wounded in action on the eastern front in December 1942 and awarded the Iron Cross for bravery under fire. On recovering from his wounds, he was recommended to lead the new commando forces that Hitler wanted to create; hence his mission to rescue Mussolini.

Later, during the "Battle of the Bulge" in December 1944, Skorzeny's commando's disguised as American soldiers, spread misinformation that there was a raid on Paris to kill or capture General Eisenhower. The Americans believed it, causing an irate Eisenhower to be confined to his headquarters for several weeks.

Near the end of the war, Hitler awarded Skorzeny one of Germany's highest military honours, the Oak Leaves to the Knight's Cross and promoted him to Colonel. Skorzeny surrendered to the Allies in May 1945 and was held as a prisoner of war for more than two years before being tried as a war criminal for his actions during the Ardennes offensive. However, he was acquitted when a senior British officer testified in his defence that Allied Commando forces had also fought in enemy uniforms.

He settled in Spain and resumed his prewar occupation as an engineer. Later, he worked as a consultant to President Nasser in Egypt and later still to the Argentine dictator Juan Peron. Skorzeny died a multi-millionaire in Madrid in 1975.

~SEVENTEEN~
BATTLESHIP TIRPITZ

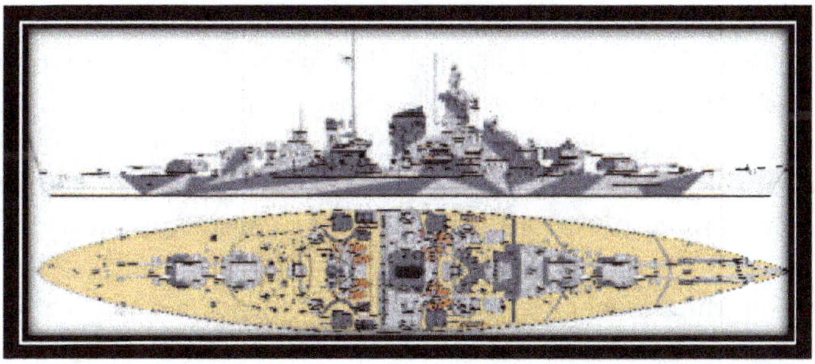

The fifty two thousand ton German battleship Tirpitz when commissioned in 1941, was the largest battleship in the world, with eight fifteen inch guns and a speed of thirty one knots. Tirpitz was named after Admiral Tirpitz who was the driving force that made Imperial Germany's High Seas Fleet a major naval power. Bismarck was the "Iron Chancellor" who unified Germany. Ironically, both men represented the beginning and end of the German Empire. Bismarck created the Empire in 1871 whilst Admiral Tirpitz led it into a naval race, resulting in war with Great Britain, and eventual defeat.

Following the sinking of the Bismarck after the action with Hood and Prince of Wales, Tirpitz was deployed in 1942 to Norwegian waters where she remained for the rest of her career, principally to guard against a British invasion that was Hitler's fantasy and be a constant menace to supply lines. From Norway, Tirpitz could attack Allied convoys to the Soviet Union or make a break for the Atlantic, as did Bismarck. Her presence alone, constituted a strategic threat as she was in effect a "fleet in being".

This concept of a "fleet in being" occurs when a naval force is smaller than its adversary and therefore unwilling to engage in full fleet combat, but is strong enough to cause concern to the superior naval power. This is due to its capacity to emerge and disrupt sea routes or catch and overwhelm part of the superior power's naval forces.

She had a crew of two thousand five hundred men and was very similar to her sister ship Bismarck, only slightly larger. She completely outclassed the older British battleships and even the new King George V class was not so well armoured nor so fast, although they did fire a heavier broadside. If these two giant German ships, namely Bismarck and Tirpitz together with other German surface units had been allowed to roam the Atlantic, the lifeline from Canada and the USA would certainly have been cut. Churchill wrote that the destruction, or even the crippling, of these ships was top priority, and their elimination would alter the entire naval situation worldwide.

Though Tirpitz never at any stage came into contact with Allied shipping, her great power caused the Royal Navy to divert badly needed resources and generated intense efforts to put her out of action. Considerable forces had to be maintained at Scapa Flow that could be otherwise employed in vital theatres of operations. The Royal Navy attacked the battleship on a number of occasions with midget submarines plus carrier based aircraft and repeated attempts were also made by RAF Bomber Command to sink her.

Tirpitz only fired her main guns offensively once during a 1943 raid on allied shore facilities at Spitzbergen. She did, however, represent the last real Atlantic surface threat faced by the Allies. In March 1942 Tirpitz left Trondheim to attack a Russian bound convoy. She did not go undetected; the British submarine Seawolf, stationed off the coast, sighted and reported Tirpitz's movements; resulting in the Home Fleet calling up three battleships and the aircraft carrier Victorious. German radio intelligence intercepted this information and Tirpitz was recalled. But en route to base she was attacked by torpedo aircraft from the carrier. In a furious battle lasting only eight minutes the Tirpitz escaped all the torpedoes and shot down seven of her attackers. Unscathed, she steamed back into the fjord, a lucky escape from the fierce and resolute British attack. Consequently Hitler ordered that warships were in future only to leave their anchorage when there was no danger of a British aircraft carrier being involved.

Perhaps the most powerful and tragic demonstration of the influence of this giant battleship came in the summer of 1942, when the mere threat of her being at sea was the direct cause of the dispersal and near-annihilation of the ill-fated Convoy PQ-17.

In September 1943 a British midget submarine raid did cause serious damage. And in February 1944, while she was still under repair, the German battleship was the target of an unsuccessful raid by Soviet bombers. In April 1944 as her repairs were being completed, Tirpitz was attacked by British carrier-based planes, receiving hits that caused major damage. Further repairs lasted until June, and she was again attacked by the Fleet Air Arm. In mid-September, she was hit with bombs dropped by RAF heavy bombers. Finally, on 12th November 1944, Tirpitz was hit by several "Tall Boy" heavy bombs, causing massive damage. She listed heavily, her magazine exploded and she met her end, taking a thousand seamen with her.

Besides tying down enormous British resources, the fear of the Tirpitz breaking out to the Atlantic prompted one of the most daring raids of the Second World War, the attack upon the heavily-defended French port of St. Nazaire in March 1942. For if the Tirpitz had broken out into the Atlantic, it was most likely that she would seek shelter in a base on the west coast of France, and the only dry dock capable of accommodating her was the great 'Normandie' dock at St. Nazaire. Therefore, the Admiralty decided to put this dock out of action.

The overall objective of the raid on St. Nazaire was successfully achieved, aerial reconnaissance showed that the gates of the lock were utterly demolished, and docking the Tirpitz would be impossible. Crippling the battleship itself, however, was still the chief priority, and three days after the St .Nazaire raid thirty three bombers attacked the Tirpitz at Trondheim. But neither this raid, nor two more strikes scored any hits. Bad weather and an efficient smokescreen shrouded the target as it lay in the narrow fjord. The aircraft attacked in two waves; the first attack by high-flying Lancaster's followed by low level Halifax's. Twenty, two ton bombs were dropped, none of which hit, and five bombers were shot down.

The battleship was actually extremely difficult to attack. She was secured in a narrow fjord, covered with camouflage nets and high mountains on either side. On top of this air-raid warnings set off an elaborate smoke screen system to rapidly obscure the Tirpitz. On the following night there was a repeat attack by thirty bombers which was again foiled by the German smoke screen. Bombs and mines were dropped around the obscured battleship, but again without success, for the loss of two bombers.

Within a month Bomber Command managed to make three attacks, a not inconsiderable achievement in view of the atrocious weather conditions. Twelve aircraft out of one hundred and seven were lost without inflicting any damage whatsoever. Something else was required but before an alternative could be planned the Tirpitz was to achieve her greatest indirect success.

In June 1942 Tirpitz and other German capital warships targeted convoy PQ-17 consisting of thirty four merchantmen escorted by six destroyers and two cruisers. British reconnaissance aircraft sighted and reported that a German battle squadron had put to sea. This resulted in a disastrous and much debated decision by the Admiralty. First Sea Lord, Admiral Sir Dudley Pound, realising that the Home Fleet was hundreds of miles away and that the cruisers and destroyers escorting the convoy were no match for their opponents, ordered the escort to withdraw while the merchantmen were told to scatter. The unprotected merchantmen were left to attempt to reach their destination independently. German naval command intercepted these reports and ordered the battle squadron to discontinue the operation and return to base. Leaving the merchantmen to be picked off by U Boats and land based aircraft.

The Tirpitz battle squadron played no part in the actual slaughter, their job was done as soon as the convoy scattered. The Admiralty panic order gave the German naval and air forces an unbelievable chance of easy plunder which they readily seized. In the next three days Allied merchant ships were sunk by bombs or torpedoes. Only ten ships eventually reached Russia. The mere threat of the surface attack had worked; without firing a gun, and without even getting within five hundred kilometers of PQ 17, Tirpitz had achieved one of the most outstanding naval successes of the war.

An attempt by the British to put the Tirpitz out of action using human torpedoes known as Chariots took place in October 1942. A Norwegian, Leif Larsen, who had escaped from Norway to Britain, was put in charge of the operation. A fishing boat was used to tow the chariots until they were close to the target. On the morning of 26th October they sailed for Norway, but on 29th October the generator to recharge the Chariots' batteries broke down.

The operation continued in the hope that the Chariot batteries were sufficiently charged. But when the Chariots got undetected to within sixteen kilometers of their target, the batteries went flat. The disappointed crew scuttled the fishing boat and Chariots and made their way to Sweden. In September 1943 a German battle squadron consisting of Tirpitz, Scharnhorst and nine destroyers attacked the Allied base on Spitzbergen. Tirpitz and Scharnhorst opened fire with their main armament and German destroyers ran inshore with landing parties. Some prisoners were taken, a supply dump and wireless station wrecked before the German ships returned safely to base. For the only time in her existence Tirpitz had fired her main armament offensively. Although her crew were not to know it, Tirpitz had carried out her last operation, for in the fourteen months remaining to her, she was to come under relentless attacks. The British decided to use midget submarines known as X-craft towed by submarines in an attempt to cripple or even sink the Tirpitz. Specialised training was carried out in Scotland during the summer of 1943.

The X-craft would be manned by a transit crew and when they were close to the target they were replaced by an operational crew. The submarines towing their X craft left their base in Scotland. Three had their tow lines parted and were separated for over thirty six hours before two were found again and continued the passage. But one was lost without trace with her transit crew. Due to technical problems another was forced to scuttle the craft and the crew taken on board the submarine.

Throughout 18th September the submarines continued to tow the four remaining X-craft and on arriving at their rendezvous point the operational crews were transferred. They successfully manoeuvered through a minefield on the surface and then dived at dawn to continue through the fjord for the final run in to their target that was sheltered behind anti torpedo nets only six km's ahead. Two of them after overcoming this hazard reached the Tirpitz. One of the X craft however, malfunctioned and broke surface. She was seen from the deck of Tirpitz, mistakenly identified as a porpoise and disregarded. She was not so fortunate a second time when, after the crew attached both charges, she again broke surface and was identified correctly.

The X craft met with a fusillade of small arms fire and hand grenades; escape was impossible. The crew scuttled the craft and surrendered to a German picket-boat and taken aboard the Tirpitz. Meanwhile two other X craft had attached their charges to the Tirpitz. The fourth X craft had become entangled in the antisubmarine nets, but eventually broke clear and attached the charges to the Tirpitz. This craft then experienced technical problems, surfaced and was forced to surrender.

German divers checked the hull of Tirpitz for limpet mines and a wire was being drawn along the battleship's hull when two violent explosions caused the Tirpitz to leap upwards several meters. All the lighting circuits and much of the power supply were put out of action and the ship settled down with a list to port. The attack had been a tremendous success. For the loss of only nine men killed and six men captured the battleship had been severely damaged; Tirpitz's main engines were put out of action and the after- turrets damaged. German casualties were one man killed and forty wounded.

Of the six X-craft which set out none returned but the six men taken prisoner came safely home after the war. Both Lieutenants Place and Cameron were awarded the Victoria Cross for the successful placing of their charges. Sub-Lieutenants Lorimer, Kendall and Aitken received the Distinguished Service Order and crewman Goddard the Conspicuous Gallantry Medal. In the German archives captured after the war was a report that "as a result of the successful midget submarine attack the battleship Tirpitz has been put out of action for six months". In fact, she did not move from her anchorage until April, 1944, after extensive repair work had been carried out. She was then moved to Tromsø Fjord.

The loss of the Battle Cruiser Gneisenau, bombed in dry dock in Germany after the Channel dash, had dissuaded Hitler from risking the return to Germany of Tirpitz and there was no dry dock big enough to take her in Norway. The British Admiralty decided that the Fleet Air Arm should attack Tirpitz as soon as the aircraft carrier Victorious had completed her refit trials. The capital ships of the Home Fleet consisting of Duke of York, Anson, Victorious, and Belfast, together with other carriers and destroyers steamed to Norway.

By dawn on 3rd April 1944 the still undetected combined force was in position. The first of the Corsairs took off from Victorious, followed by Barracudas, Wildcats and Hellcats from the other carriers. As the strike was approaching Tirpitz the smoke screen ashore was activated. The Corsairs remained at three thousand meters to cover the Barracudas against counter-attack. The Wildcats and Hellcats came in low over the hills, strafing the battleship with machine-gun fire as the Barracudas began their bombing dives. The bombs were meant to be dropped from a height above one thousand meters to ensure penetration of the armoured deck but in their enthusiasm and determination to achieve accuracy most of the pilots dived much lower than this. Six direct hits were claimed plus three probable hits and the battleship's upper deck was left bloody with the wounded.

As the first strike was attacking the second strike made for Tirpitz. The Hellcats attacked the battleship's anti-aircraft positions with the Wildcats strafing the unprotected bridge and upper deck before the Barracudas came in to the attack. Eight definite and five probable hits being claimed, but the 725 kilogram bomb which hit the bows failed to explode. By 08h 00 all the aircraft had landed back on their carriers except for two Barracudas shot down. In just two minutes the Fleet Air Arm had ruined much of the six months of repair work and had done incalculable harm to morale. The upper deck was a shambles and the casualty list formidable. One hundred and twenty two men had been killed and three hundred and sixteen wounded including the captain, many of them by the machine gun fire from the fighters. It was not until June 1944 that Tirpitz could recommence trials.

On 17th July 1944, another attack was mounted by the Fleet Air Arm. Forty four Barracudas, loaded with 450 kilogram and 225 kilogram bombs were escorted by eighteen Corsairs and fifteen Hellcats. Tirpitz received warning of the approach of a large formation of aircraft, and the battleship was totally obscured by a smoke screen. No hits were observed and one Barracuda and one Corsair were lost.

The Tirpitz was only finally destroyed when British inventor, Sir Barnes Wallis, built a special bomb for the job. The same man who had previously developed the bouncing bomb used in the Dambusters Raid. In 1944 he devised the "Tallboy," a_five and a half ton bomb capable of piercing the Tirpitz's armoured deck. A most devastating bomb designed for accurate flight and great penetration.

As the war progressed both the technique and technology of the Royal Air Force had increased enormously. The combination of the genius of Barnes Wallis together with the inspiring leadership of Wing Commander Guy Gibson led to the creation of the elite 617 squadron. The most determined and skilled squadron in Bomber Command.

The Tirpitz was attacked again on 29th October, thirty seven Lancaster's were dispatched from Lossiemouth, Scotland. The removal of the mid-upper gun turrets and the installation of extra fuel tanks meant the Tirpitz could now be reached directly from land. Thirty two aircraft released Tallboys but no direct hits were scored. Early in the morning of 12th November Thirty two Lancaster's took off on the final raid against Tirpitz, aptly named Operation "Catechism". As the Lancaster's came over the mountain range they were met with intense anti-aircraft fire from Tirpitz, shore batteries and flak ships, but the battleship herself lay clearly visible. Astonishingly there was no smoke-screen or fighter cover. The Lancaster's were equipped with the new and deadly Mark XIV bombsight and conditions were ideal.

Twenty nine bombs were dropped. The first bombs narrowly missed the target, but then a great yellow flash burst on the foredeck and the Tirpitz was seen to tremble as it was hit by another two Tallboys. A column of steam and smoke shot up to about one hundred meters and within a few minutes the ship started to list badly. Flames and smoke belched out of Tirpitz's stricken hull, and she immediately took on a list of 30 degrees to port.

No ship, not even Tirpitz, could withstand direct hits by such great bombs. Aboard the stricken vessel counter flooding had been ordered to correct the list, but the already slim chances of saving the ship were quickly crushed by another bomb spectacularly obliterating one of her main turrets, while further near misses shook the ship violently.

Her end came very suddenly. The first two direct hits had pierced her vitals and while she listed still further, her after-magazines blew up with a tremendous explosion, causing her immediately to roll over. As the last of the Lancaster's flew homewards, the crews could see the Lone Queen of the North, was dead. The lack of fighters and smokescreen proved to be disastrous.

The eight German fighters which took off had been misdirected and recalled before they reached the Tirpitz. Investigation into the failure of the smokescreen revealed that it had not been operational due to repairs. More than one thousand men were trapped when she heeled over so suddenly. Over eighty of these picked their way through the various decks to the upturned bottom of the hull, upon which they knocked furiously, attracting the attention of those outside. A hole was cut in the steel hull through which they escaped, thirty hours after the battleship turned turtle. Other knockings were heard too, but before the unfortunate men could be reached, the waters of the fjord filled their compartments. The rescue teams heard the sailors chanting 'Deutschland Uber Alles' before silence settled upon the hull.

Tirpitz had perhaps the most inglorious and tragic career of any warship in the Second World War. She never saw action with an Allied convoy or battle fleet and activities were limited to scurrying from one Norwegian fjord to another, and making furtive excursions which ended as soon as any danger approached. She was the last heavy warship left to the German navy by 1944; but for the British, she was too great a threat to be ignored.

Not for Tirpitz the fate of the Bismarck, a brief moment of glory and then a valiant fight against the odds, and she was always unable, apart from the minor affair at Spitzbergen, to use her strength in action because of Hitler's fear of losing capital ships. Her destruction was thus piecemeal and her end has an air of tragic inevitability, for though she sank quickly enough when the Lancaster's bombs ripped her apart; she had in reality been wasting away for months before that.

And yet, some success must be accredited to her. She had brilliantly achieved the aim of a 'fleet in being' by tying down far greater enemy forces. Whilst sea worthy she had presented a permanent threat to the Atlantic and Arctic sea-routes, and a tremendous annoyance to the Royal Navy, which desperately needed to send the battleships and aircraft-carriers of the Home Fleet to more active theatres, yet dare not while Tirpitz floated.

Little wonder, then, that all possible means were used to destroy her and her epitaph, if one is to be written, should be, "she could not be left alone".

~EIGHTEEN~
TRAGEDY AT SLAPTON SANDS

In the early hours of the 28th April 1944 a convoy of eight LST's (Landing Ship Tanks), full of American servicemen and equipment was nearing Slapton Sands on the south coast of England. Their purpose - to take part in Operation Tiger, the military code name for a D Day exercise that was due to take place in a few weeks' time at Normandy.

The success of D DAY, to attack Hitler's much vaunted Atlantic Wall was the largest amphibious assault in history, and would depend on precise coordination and complete secrecy. The sacrifice of those who lost their lives on the beaches of Normandy is well documented. However, what is largely unknown until recently is the slaughter at a practice landing whose mission was to prepare for the invasion. Their sacrifice possibly ranks alongside that of Dieppe in helping to ensure the success of D Day.

Operation Tiger at Slapton Sands was one of several military exercises to simulate the landings at Utah Beach. But during this exercise hundreds of American servicemen perished mainly due to confusion and negligence. It was also one of the military's best kept secrets and not entirely revealed until military authorities finally acknowledged the event thirty years later when the Freedom of Information Act was passed in 1974.

It happened just as the assault ships and their escort vessel were manoeuvering into position to start their practice landing. Suddenly, out of the darkness a group of German E boats appeared in attack formation. In the resulting chaos and carnage two LST's were sunk and two badly damaged. The beaches became littered with dead bodies with more being brought in on each tide. When the waters of the English Channel finally ceased to wash bloated corpses ashore, the toll of the dead and missing were counted. It turned out to be the most costly training incident involving U.S. forces during World War Two. Seven hundred and forty nine American servicemen had lost their lives. Three times more casualties than when 'Utah' beach was stormed on D-Day. Due to concerns over possible information leaks just prior to the real invasion, the whole affair was hushed up and many of the dead were hastily buried in unmarked graves.

Generals Bradley and Eisenhower issued orders that the families of the dead were not to be informed of how the men died. They made sure that details of the tragedy would be buried with the troops. Under pain of court martial the strictest secrecy was imposed on all who knew of the disaster. This included medical staff that treated the wounded survivors.

Security of the entire D Day operation was considered compromised and perhaps of more significance, at a high military level, confidence was badly shaken in the ability of the British and Americans to successfully cooperate.

Meticulous planning for the cross channel invasion began in the spring of 1942 mainly due to Russian pressure for a second front. But, where was that second front to be? The Americans wanted to launch a cross channel invasion immediately. But British commanders haunted by their First World War experiences considered this to be too risky. This was somewhat justified in August 1942 when the disastrous raid on Dieppe proved that an attempt on Fortress Europe would fail due to the lack of adequate preparation and strength.

The British gradually brought the Americans round to their way of thinking. Both parties eventually agreed that it would be impossible to achieve the required troops, weapons and equipment necessary to gain the overwhelming strength needed to guarantee success. At this time there was the calamity of Rommel capturing Tobruk whilst Churchill was in Washington meeting with Roosevelt. Priority was immediately switched from a cross channel operation to that of invading French North Africa to relieve pressure on the British.

This infuriated some American military chiefs, particularly General Marshall and Admiral King who had been insisting all along to prosecute the war against Japan first not Germany first. They now put their case stronger than ever, stating the British were only half hearted regarding a cross channel operation. But Roosevelt had committed himself to Churchill and dissuaded Marshall and King. There was also still the real risk of upsetting Stalin whom it was feared might come to a separate peace with Hitler.

Nevertheless an agreement was reached, and Operation Torch, the Allied invasion of North Africa took place in November 1942, culminating in the total surrender of all Axis forces in Africa in May 1943. The British and American views once more diverged sharply. The British were keen to exploit what they considered to be the "soft underbelly", of Europe whilst the Americans were totally opposed to becoming "bogged down" in the Mediterranean. Tension between the Allies was further increased when the topic of a supreme commander was broached.

General Sir Frederick Morgan had since 1942 supervised the planning of the Normandy invasion. He headed up COSSAC – Chief of Staff Supreme Allied Command. By late 1943 Churchill had come round to accepting it would have to be an American in overall command. He was in favour, as was just about everybody else, of General Marshall the US Army chief of staff. The appointment of a supreme commander became top priority, particularly as hundreds of thousands of American troops were pouring into Britain in preparation for the invasion. But still the politicians dithered, mainly due to the sensibilities between Limeys and Yanks; stigmatized with overpaid, oversexed and over here.

At the Cairo conference in December 1943, Roosevelt declared that he could not spare Marshall. The Supreme Commander would be Eisenhower with RAF Chief Tedder as his deputy. Various other appointments were made and with SHAEF established, preparations for a landing in six months time began. The distrust between the American and British commanders makes one wonder how the invasion ever got launched. That it did is an enormous tribute to the patience of Eisenhower and Churchill stepping in time after time to intercede in bitter quarrels.

Operation Overlord required coordination, leadership and dedication to duty under the supreme command of General Eisenhower. This was not the view of the British who continued to hold the American military in a sort of contempt. Field Marshall Sir Alan Brooke kept reminding his associates that he fully expected disaster, complaining that Eisenhower had never even commanded a battalion in action. Roosevelt, in a conciliatory gesture, approved to the temporary appointment of Montgomery as overall land commander of the invasion forces, until Eisenhower established his headquarters in France.

On both sides of the channel, preparations gained momentum. The appointment of Eisenhower as supreme commander made it apparent to the Germans that an invasion was imminent. This convinced Hitler of the need to pour enormous resources into strengthening the Atlantic Wall. He convened a meeting of his top advisors including Göring, Speer and von Rundstedt. It was decided there would be fifteen thousand concrete strong points manned by three hundred thousand troops. Work was concentrated on the Calais area where four times the resources of Normandy was allotted. Hitler then appointed Field Marshall Rommel to inspect the defences and to take command of Army Group B.

The major American deficiency going into Normandy was their tanks. They had concentrated production on the Sherman which by German standards was totally obsolete. German 88 mm guns mounted on Tiger tanks could knock out a Sherman before it even got within firing range. But Allied superior air power was seen as the solution to this problem.

Another factor of considerable importance was the almost complete lack of German information about Allied plans and dispositions. This was further aggravated by the inter-service rivalry within the

German High Command. The result was that the Germans had no real concept of the strength of Allied forces, to the extent that Hitler even considered transferring troops from the West to the Eastern front. The key to Allied success was to surprise the Germans and keep them off balance long enough to allow for the entire invasion force to land and be supplied.

Meanwhile, Eisenhower soon came to realise that the plan proposed by COSSAC would have to be broadened from three divisions to five. More troops, more beaches and more LST's were required. The Americans now required five thousand tons of equipment per day and the British seven thousand. This meant that the invasion could not possibly take place in May 1944 and had to be postponed by a month. The Russians were told of the date of the invasion and agreed to stage an offensive on the eastern front at the same time.

Ultimately more than eight hundred thousand British and Canadian troops and one million three hundred American troops were assembled to take part in the great battle to liberate Europe. Another key to the success of the Normandy landings was deception operations that were building up on the southeastern coast of England at the same time as the actual landing forces were concentrating in the southwest.

General George Patton, well known and much feared by the Germans sent out a variety of messages from his dummy headquarters on the east coast of England. This was intended and succeeded in fooling the Germans into believing that the main invasion would come at the Calais area.

Amphibious operations, according to military theorists favour the defenders by about five to one. The defenders also had the advantage of moving troops from several areas to overwhelm the attackers and throw them back into the sea. Tanks, big guns and heavy trucks were at serious risk on the landing beaches. The task would be greatly simplified if they and thousands of tons of other supplies were able to dock in a captured port and be disembarked in an orderly manner. But there were no ports; the Germans had turned them all into fortresses protected by heavy batteries, and they had been wired for demolition if the Allies approached, it would take months to restore them.

Realising the difficulties of capturing a harbour to offload the supplies needed to maintain momentum the planners came up with the idea of the Mulberry Harbours. If a port cannot be captured then an artificial one must be taken along. Due mainly to the initiative of Churchill it was decided to take two enormous artificial harbours with them to France. The Mulberries consisted of concrete caissons to be sunk off the Normandy coast and floating piers able to rise and fall with the extreme Normandy tides.

All was now ready for the troops to undergo one last big dress rehearsal prior to the invasion. Therefore, the practice assault on Slapton Sands was planned to be as realistic as possible. Suitable training beaches had been identified that would simulate landings on the Normandy coast. Slapton Sands was selected because it had characteristics similar to "Omaha" and "Utah" beaches. Other beaches were also selected to the east of Portsmouth to simulate landings on the British zones, namely, Gold, Juno and Sword.

The continued shortage of LST's was a major headache for Allied planners. The LST was the most valuable vessel afloat for Normandy, without which the invasion could never have taken place. Britain and Canada built some but the vast majority constructed in the United States. The Landing Ship Tank was developed in response to a need to transport armoured and infantry troops by sea.

These shallow draft flat bottomed assault ships of four and a half thousand tons could steam at 10 knots and were capable of carrying several hundred men, vehicles and tanks. Measuring one hundred metres by ten metres LST's could be described as the world's first mass produced "roll on roll off" ferries. Over flat beaches such as Normandy an LST could operate in about one metre of water. They were constructed so that the bow opened up, a ramp went down and tanks and trucks could literally be driven ashore.

The Allied Command was certain that the Germans were aware of general invasion intentions. Therefore, particular attention was paid to guarding information on the three areas of vital surprise. That is the date of attack, the place of attack and the strength of attack.

The decision to invade Normandy was so secret that an elaborate system called BIGOT was set up to watch for any possible leaks. The unique classification of BIGOT was accorded to any documents on the subject. Inevitably people privy to the information became known as bigots! But despite of all precautions, some near disasters did occur. For example an envelope Marked BIGOT was found opened in a Chicago mailing office and contained vital information of D Day. FBI agents rushed to the scene to discover that it was addressed to a person of German descent. However, the investigation revealed that a genuine mistake had been made by the mailing clerk. Nevertheless close surveillance was maintained on all suspects until after D Day.

Eisenhower received a nasty shock when it was reported that his old friend General Miller got drunk at a dinner held at Claridges hotel in London and was heard by several guests to name the date of invasion. Eisenhower reduced Miller to the rank of Colonel and sent him back to the States in disgrace. The threats to security continued unabated. A British staff officer left a brief case containing top secret documents in a London taxi that fortunately was turned in by the cabbie.

In the spring of 1944, the Germans maintained fifty eight divisions in France and the Low Countries poised to face an invasion. They had a good idea that it would be launched in late spring or early summer. But the vital question was where?

Most high ranking German officers were convinced that it would come in the Pas de Calais area. Hitler almost alone suggested Normandy and ordered that coastline to be reinforced.

In late 1943, the British Government evacuated approximately three thousand local residents from their farms and villages in the Slapton Sands area and the American Army moved in. Tens of thousands of American troops were billeted around this area to undergo training for the imminent landings. The surrounding area was converted to resemble what they would meet on the other side or the far shore as it was called. Interlocking fields of fire were established as were pillboxes, mines and beach obstacles. Provisions were also made for the collection of prisoners of war and burial parties.

Over the weeks that were to follow the Americans made many training attacks on the beaches until all was ready for the full scale rehearsal, Exercise Tiger, under the command of US Admiral Moon. So vital were the exercises that live ammunition was used to make it as realistic as possible and accustom the troops to what they were about to experience.

To protect the LST's at the practice landing at Slapton Sands, the Royal Navy Commander in Chief, based in Plymouth, ordered naval patrols off the landing beaches. A motor torpedo boat patrol was also sent to watch Cherbourg, the main port where German E-boats armed with torpedoes and a top speed of forty knots were based.

The assault convoy carrying vehicles and combat engineers left Plymouth at 9.45 pm on the night of 27th April 1944 and sailed at a speed of five knots in a single row, keeping a distance of about four hundred metres. General Eisenhower, Air Marshall Tedder, General Montgomery and General Bradley plus other members of the SHAEF contingent split up so that all these high official observers would not be on the same vessel.

At 06h 20 it was reported to Admiral Moon that some of the LST's were behind schedule and it was requested that H Hour be delayed for an hour to which he approved. Due to a typing error in orders, the LST's were on a different radio frequency from the escort and naval headquarters ashore. When one of the picket ships in the Channel spotted German E boats soon after midnight, a report quickly reached the British escort but not the LST's. The escort commander assumed the LST's had received the same report and therefore, made no effort to contact them.

Escort and convoy may well have been on opposite sides of the Atlantic as far as the radio communication was concerned. The army officers of the second wave did not receive the order and continued to land troops at the appointed time on beaches designated to be fired on with live ammunition. Everything from then on seemed to go wrong. If this was how things were going to be on D Day there would be wholesale slaughter.

Meanwhile across the English Channel at Cherbourg, E Boats were preparing for a patrol run along the southwest English coast. Previously, E boats operating out of the Channel Islands had successfully attacked two Allied convoys and sunk five vessels, including a Norwegian destroyer. As the amount of slow moving traffic in the Channel increased due to training exercises so too did the targets of opportunity for the fast E-boats. These successes made them increasingly bold and they were confident of a good strike, as Slapton Sands would so tragically confirm.

E-boats were thirty five metres in length with a crew of twenty one. They were powered by triple shaft Daimler-Benz diesel engines and equipped with twin 21 inch fixed forward facing torpedo tubes and two either 20mm or 37mm canons. As the convoy approached Slapton Sands, British shore batteries defending the harbour observed silhouettes of the E-boats but were instructed to hold fire so as not to alert the Germans of the exercise.

The German naval commander at Cherbourg was alerted by the greatly increased naval activity in the Slapton Sands area, and ordered E boats to that vicinity. The E Boat commander Captain Rudolf Petersen, maintaining radio silence to avoid detection soon came into visual contact with the LST convoy and swiftly positioned for a torpedo attack.

All the LST'S were brand new arrivals in the UK with brand new skippers, mainly ninety day wonders commanding brand new crews. They had never operated together as a group and steamed slowly on completely unaware of the danger lurking on the flanks

The convoy was caught surprised even though Allied intelligence had warned about possible danger from German E Boats. By this stage of the war, there was no threat from German Capital ships which, had been generally neutralised. Much more serious, however, was the threat from the E-boats bearing down on Slapton Sands.

Within minutes of the attack torpedoes hit three of the LST's, two burst into flames, rolled over and sank. Trapped below decks hundreds of servicemen went down with the ships. There was no time to launch lifeboats, causing many to leap into the sea and weighed down by waterlogged coats soon drowned. Others, whilst waiting to be rescued succumbed to hypothermia in the cold water.

Troops unused to being at sea panicked and put on their lifebelts incorrectly; as a result, when they jumped overboard, the weight of their combat packs flipped them onto their backs, pushing their heads underwater to drown. Intermittent firing was seen and heard but could not be identified in the darkness and confusion. Further torpedoes were fired by the E Boats and found their mark as another LST burst into flame.

Some LST's were initially more fortunate; the E Boat skipper thought they were Destroyers and set his torpedoes to run too deep before realising they were shallow draft vessels. While correcting this error, E Boat 40 mm guns opened fire to inflict further damage. The deck was a terrifying experience for those men who made it to the top. They were surrounded by blazing fires and came under machine gun fire from the E Boats. Those who did manage to get into a lifeboat still had a long ordeal ahead whilst awaiting rescue. The E-boat commander decided to terminate the action and made for Cherbourg at high speed to escape without loss.

British aircraft were alerted and made several attacks, damaging one of them. All nine E Boats, however, made it back to their French base. It was an amazing victory. As the LST convoy reformed and headed for the beach, the escort vessel took station astern, zigzagging. The captain did not go to the assistance of the flaming LST's. As the only escort present he had strict orders to remain with the convoy.

When the surviving LSTs landed on the beaches, the blunders continued and many Americans were killed by friendly fire from the British artillery that had started shelling with live ammunition. On the beaches a white tape was placed indicating where the troops should not proceed beyond until the live firing had finished. But they kept going straight through the white tape line and were killed or wounded.

By the time Admiral Moon learned of the E Boat attack and the sinking of the LST's it was too late for him to take any effective action. The question very much on everyone's mind was how much did the Germans know about Operation Tiger? Had the E Boats simply blundered into the convoy on a routine patrol, or, had enemy agents reported that a practice landing for an invasion of France was being conducted? Of vital importance, had the enemy taken prisoners who had knowledge of Operation Overlord, they may be in possession of all the important facts as to the time and place of the invasion.

It seems inexplicable that Operation Tiger took take place without adequate escorts. Especially in the full knowledge that there was a potent E-boat threat in that area. Rear Admiral Moon came in for much hostile criticism from Allied High Command. In assessing the exercise it was concluded that in war lives are lost, sometimes apparently meaninglessly, and the disaster at Slapton Sands was only part of what may be expected for real on D Day. It should serve as a lesson and greater naval protection was to be provided for the invasion forces as they approached the Normandy beaches. But the fact emerged that Admiral Moon had made a serious error in postponing the exercise for an hour. Some units got word of the change and some did not. The whole landing procedure was sloppy. If that were to happen in Normandy, the result would be catastrophic. Enormous casualties were now expected on D Day and General Eisenhower feared that the whole operation might fail. Changes had to be made.

An official investigation was ordered and two major factors were revealed that contributed to the tragedy — a lack of escort vessels and an error in radio frequencies. Once the assault waves landed on the beaches the most important task was traffic direction. It was essential that equipment and supplies be moved fast; otherwise they would present an easy target. The Germans could be expected to pull themselves together and launch counterattacks. Slapton Sands observers found there were too many officers with little leadership skills, battle discipline, and lacking in aggression.

The problem on this simulated D Day exercise was organisation, or the lack of it. Troops were allowed to walk about without knowing what to do. Combat vehicles were bogged down and the beach masters failed miserably. Slapton Sands highlighted major deficiencies in American command and combined operations. The system of command was clearly established at a high level but did not function on the operational level and lacked flexibility.

The most serious damage of all, the loss of seven hundred and forty nine American servicemen in the E Boat attack was blamed on the Royal Navy in failing to provide minimal protection for the lumbering LST's. Also, the escort had led the LSTs in a formation which presented an easy target to the E-boats. The original plan was that the convoy would be accompanied by a Royal Navy Corvette and a Destroyer. It was claimed that this had proved adequate in the past. The Destroyer, however, was damaged in a collision and put into port for repair, a replacement not being immediately available. Radio frequencies were standardised. Better life drill training for troops and small craft to be available to pick up D Day casualties.

When the news of the disaster reached the Allied commanders on shore they became greatly concerned. Ten officers aboard the LST's had been closely involved in the invasion planning and knew the assigned beaches in Normandy; obviously information of vital interest to the enemy. There was a strong possibility that the plans for the invasion had been fatally compromised. If the Germans had taken them prisoner they might force them to reveal information about the invasion. The question arose, what was the purpose of the E Boat attack. Was it to take prisoners who had knowledge of Operation Overlord? If such prisoners had been taken should not SHAEF consider calling off the whole invasion? But if the invasion was postponed one more time then in all probability it would not be launched in 1944 and there would be hell to pay with the Soviets.

A vast search was undertaken and the bodies of all ten officers were recovered and the tactics of D-day remained secure. This was borne out by Bletchley Park who was watching for any immediate action from the other side. If the secrets of D Day had been revealed then a re-shuffle of enemy forces could be expected. This did not happen and German forces remained mainly in the Pas de Calais area.

But equally serious was the loss of the LST's. This meant that not a single LST was available as a reserve for D-Day. This would have dire consequences at Omaha beach. The first round in the invasion of Normandy was won by the Germans and Captain Rudolph Petersen was awarded Oak leaves to his Knights Cross.

Conversely, Rear Admiral Moon was summoned by U.S. Admiral Kirk and hauled over the coals. Three months later the death toll increased yet further. Moon never got over the disaster at Slapton Sands and because of what was officially termed *"combat-related stress," due to the ill-fated Exercise Tiger; he put a revolver in his mouth and shot himself.* Rear Admiral Moon was the highest-ranking American officer to commit suicide during World War II.

Renewed confidence returned to the Allied command in the knowledge that the secrets of D Day had not been compromised due to the action at Slapton Sands. But this was suddenly shaken just a week after the event. The ULTRA listeners at Bletchley Park picked up Hitler's new orders for the defence of Normandy. The orders called for the strengthening of this area and aerial photographs confirmed this was happening and troops were transferred from as far as Hungary to St. Mere Eglise and Carentan. Slap in the middle of 101st and 82nd airborne divisions drop zones.

The whole question of security was opened again until it was revealed that the entire flap had been caused by Hitler's so called intuition. He awakened one morning and simply decided that he ought to strengthen the Normandy area and issued orders to do so. There had been no breach of Allied security.

The major problem now faced by Eisenhower was the weather. Meteorologists forecast gloomy weather for 5th June that caused a postponement of twenty four hours. It was now a question of either scrap the invasion or go. The Supreme Allied Commander made the fateful decision and the invasion was on. It was a bold and shrewd decision. In a matter of hours the men would show what they had learned in the last few months about amphibious operations, and particularly what had been learned at Slapton Sands. H Hour was set for 06hrs 30 on the morning of 6th June. There would be no postponement; the lessons at Slapton Sands had been noted.

By war's end, the tragedy off Slapton Sands, like many another wartime events involving high loss of life, received little attention. In 1954, ten years after D-Day, U.S. Army authorities unveiled a monument at Slapton Sands honouring the people of the region; "Who generously left their homes and their lands to provide a battle practice area for the successful assault on Normandy in June 1944".

In 1968 a former British policeman, Kenneth Small, retired to a village nearby Slapton Sands. He took long walks along the beach and began to find relics of war such as cartridge shells, buttons and fragments from uniforms. He spoke with many people who had long lived in the region, and learned of the heavy loss of life in Exercise Tiger. He raised the question - Why, was there no memorial to those who had died? Was there an official cover-up?

There was the monument the U.S. Army had erected to the civilians, but there was no mention of the dead American servicemen. From local fishermen; he learned of a U.S. Sherman tank that lay beneath the waters just offshore. The tank was salvaged and placed on a plinth as a memorial to those Americans who had died and dedicated in a ceremony on the 40th anniversary of D-Day. This tank commemorates those killed during exercise 'Tiger', many of whom may still lie buried in unmarked graves to this day.

In the bigger picture Slapton Sands represents only a small fraction of Normandy in particular and the Second World War in general. However, this does not detract from the courage and sacrifice of those who gave their lives so that victory was made possible.

In the final analysis, Operation Tiger was not useless and the deaths of those soldiers and sailors aboard the LST's were not meaningless. Many lessons were learned at Slapton Sands and put to good use when the real invasion was launched. The men who died on that chilly April morning contributed a great deal to the success of the landings that were carried out in France just six weeks later. If nothing else Operation Tiger convinced all involved that they had to be flexible and ready for the unexpected. The troops had learned to improvise and the handful of casualties at Utah beach proved this to be the case.

D-Day was the culmination of over two years of meticulous planning, preparation and massive effort by the Western Allies. In an operation that involved the formation of the largest invasion fleet in history, the massed Allied armies successfully landed in Normandy to punch a hole in Hitler's seemingly impregnable Atlantic Wall.

If in the words of Winston Churchill the Battle of Alamein was the end of the beginning, then surely D-Day was the beginning of the end for overstretched and visibly crumbling, Nazi Germany. Hitler must be held accountable for placing the Third Reich in a virtually hopeless position, and Normandy was his death blow.

Altogether one hundred and fifty six thousand men came ashore or dropped from the sky that day, and although each man knew that this was the biggest amphibious invasion ever, few had much time to reflect on the fact. That was for the military leaders watching anxiously from the big ships lying offshore, or following the maps at their headquarters back in England. For most of the men on the Normandy beaches it was enough, at the end of the day, to be ashore and still be alive.

In the next two months the Allies landed over two million men, five hundred thousand vehicles and three million tons of stores that were necessary to drive the Germans out of occupied Western Europe. The tragedy of Slapton Sands certainly played a major role.

~NINETEEN~
D – DAY: A DIFFERENT PERSPECTIVE

D-Day – also known as the 'Longest Day' – or - Operation Overlord was the culmination of over two years of meticulous planning by the Allies. In an operation that involved the formation of the largest invasion fleet in history, the massed Allied armies successfully landed in Normandy to punch a hole in the seemingly impregnable walls of Hitler's 'Fortress Europe'.

Supported by paratroops, massive aerial bombing, and off shore naval bombardment, the landing troops secured their allotted beachheads on the first day. Allied success from the bullet-swept beaches, through the hedgerows of Normandy and the break out, would ultimately lead to victory. This was testimony to the courage and determination of the fighting men.

After the St. Nazaire and Dieppe raids in 1942, Hitler ordered the defences of all naval and submarine bases to be strengthened. Subsequently, coastal batteries were built at key places to protect ports and estuaries. Fortifications remained concentrated in these areas until late 1943. By then Hitler's war had started to go awry and defences were increased in all coastal areas from Norway to the Spanish frontier. It was a massive undertaking and German propaganda named it the Atlantic Wall.

In anticipation of an invasion, Rommel was assigned early in 1944 to inspect anti-invasion defences. Never a great proponent of static defence, he declared the existing coastal defences to be entirely inadequate and immediately began strengthening them. Rommel was aware of the old adage that he who defends all defends none. Nevertheless he was a military man and at this stage still loyal to Führer and Fatherland.

Organisation Todt was appointed the chief engineering group responsible for the design and construction of the wall's major fortifications. Thousands of forced labourers were used in their construction. The gigantic Atlantic Wall composed of blockhouses, minefields, and many kilometers of barbed wire, flame throwers, machine guns, and anti-tank ditches. Rommel intended to destroy Allied landing craft with underwater obstacles placed just off shore. Many radar tracking stations as well as listening stations were also constructed on what he considered the most likely landing sites.

By the time of D Day the Germans had laid almost six million mines in Northern France. Additional gun emplacements and minefields were extended inland, along roads leading away from the beaches. Slanted poles with sharpened tops, known as "Rommel's asparagus" were placed in possible glider and paratroop landing spots. The Germans maintained fifty eight divisions in the West poised to repel an invasion. However, only fifteen of these were considered first rate. They had a good idea that the invasion would be launched in late spring or early summer 1944. But the vital question was where?

Even if the Allies could not hide that an invasion was imminent, they were determined to sow as much confusion as possible for enemy intelligence. Blinded by the greatest deception operation in history, German intelligence was misled into a calamitous misinterpretation of Allied intentions. Most high ranking German officers were convinced that the invasion would be in the Pas de Calais area.

The German Commander in Chief, Field Marshall von Rundstedt had no faith in forts. He was acutely aware of the wall's weaknesses stating that it was nothing but an illusion fostered by propaganda to fool the Germans as well as the Allies. The fate of the immensely powerful Maginot Line had proved that defenses were no stronger than their weakest point. Rundstedt believed that an actual landing could not be prevented and he planned therefore, to hold in strength only key ports and the most vulnerable sections of the coast.

By these tactics he hoped to delay an Allied build up long enough to enable him to launch counter attacks and drive them back into the sea. Rommel disagreed and vehemently argued that Allied air power would disrupt Rundstedt's plan. Once Allied forces had secured a beach head they would inevitably break out. If there were to be any chance of success, the invasion would have to be defeated on the beaches. It was imperative he maintained that the maximum force should oppose the invasion on the very day of the landings. Their differing viewpoints led to a fatal compromise. The Panzer divisions were to be held well back from the coast line and only the infantry committed.

D Day, the Allied liberation of Europe came at last. The Allied armies landed in Normandy on June 6th 1944. Even as they approached the battle was already under way. Since midnight, in order to safeguard the flanks of the advancing troops, glider and paratroop forces had been fighting inland for a perilous foothold. French Resistance fighters blew up bridges, derailed trains and sabotaged telephone lines. Out at sea, the guns of the big warships threw out massive salvos, while overhead Allied fighter and bomber aircraft completely dominated the air.

Somewhat perversely, Hitler welcomed the Allied landing. He was sure that the German forces would repeat the carnage suffered by the Canadians at Dieppe. By nightfall on June 6th the much vaunted Atlantic Wall had been breached on a front of eighty kilometers. Rommel had failed to smash the invasion on the beaches. In fact the German commanders at all levels failed to react swiftly to the assault phase; probably due to communication problems caused by Allied air and naval firepower. The German complete lack of intelligence led them to believe that the main invasion would focus on the Calais region. They considered Normandy to be merely a feint. With the result that the Allies were not met on the beaches in the strength they expected, except of course at Omaha.

Two other important factors are highlighted; firstly the Allies landed in bad weather, which took the Germans by surprise and secondly the Germans were certain that the landings would take place at high tide. But the Allies went in at low tide; consequently once they had consolidated their landings, the Germans had lost the battle of the longest day. D Day was over and the battle for Normandy then began.

Once the Allied beach heads had been safely established and linked, two artificial Mulberry harbours were towed across the English Channel in segments and made operational by D+3. One was constructed at Arromanches by the British, the other at Omaha Beach by the Americans. On June 19th, severe storms interrupted the landing of supplies for several days and destroyed the Omaha Mulberry. Remains of the Mulberry's can still be seen to this day on the beach at Arromanches.

Pre-invasion plans called for the British to push inland thirty kilometers on the first day and take the critical road junction at Caen. At the opposite end of the line, American forces were supposed to cut across the Cotentin Peninsula turn north and take the great port of Cherbourg by D plus 8. Practically none of these scheduled objectives were achieved. In reality, by the middle of June, the expansion of the Normandy beachhead was way behind the timetable. Caen was still in German hands, and Allied forces were nowhere near Cherbourg. They were also contained on the way to Saint-Lo, the offensive in effect had slowed to a crawl.

There were two major reasons why the Allied advance had bogged down; there was the tenacity of the Germans plus their skillful use of the terrain. The Germans were aware that the mortal threat to their forces lay in the Caen area. Beyond Caen and towards Falaise, the country was wide open and ideally suited for tank operations; therefore, posing the threat of encirclement of all their forces in Normandy. A breakthrough in this area could spell disaster for the Germans, and under- mine their whole defensive position in the West.

Accordingly, they deployed the majority of their forces in or around the area of Caen and ferociously resisted every attempt by the British to capture the city. West of Caen, mainly in the American sector, the terrain favoured the defence, and the Germans could afford to spread their forces more thinly. This was the hedgerow country, or bocage as the French called it. It was a patchwork of thousands of small fields enclosed by almost impenetrable hedges, consisting of dense thickets of hawthorn and brambles, up to five metres in height, with a drainage ditch on either side. The Allies were fighting in what may be described as literally blind countryside, and a defenders paradise.

Faced with complete Allied air superiority, the Germans became masters of camouflage, using the dense vegetation to conceal men and vehicles. The drainage ditches were also deathtraps for tanks. They were easy marks for German antitank rocket launchers, the dreaded Panzerfausts. A tank that ventured off the road and attempted to smash through the thicket was particularly vulnerable. As it climbed the mound at the base of the hedgerow, its guns were pointed hopelessly skyward and its underbelly was exposed to fire from antitank guns.

Despite considerable Allied material superiority, the Germans kept the Allies bottled up in Normandy for nearly two months. German survival depended on stopping an Allied break out. Therefore, to the east and west they put a containing ring of steel around the Normandy beachhead. Fortunately, General Patton's deception army before the invasion had kept German attention focused on the Pas de Calais, and indeed high-quality German forces were kept in this area, away from Normandy, until July. So while the fighting raged in Normandy, the German Fifteenth Army, approximately two hundred thousand troops guarded the Calais coast against an attack that never came. When these troops were finally released late in the day, any effective counter stroke had no chance of success. By this time, the Allies had landed over two million men, five hundred thousand vehicles and three million tons of stores. St Lo and Caen finally fell and the Allies emerged from the hedgerow country. But all of this had come at a tremendous cost.

The British, Americans and Canadians had suffered one hundred and twenty two thousand casualties since the Normandy landings. But they had inflicted terrible casualties on the Germans, over one hundred and fifteen thousand were killed, wounded or missing. The cost of liberation was hideously high for French civilians also. Three thousand civilians were killed during the first twenty four hours of D Day. That is twice the number of Allied troops killed on D Day.

As the war swept through Normandy, hundreds of thousands of men, women and children found themselves caught between the opposing forces. While Allied planes and artillery relentlessly bombarded towns and villages that the enemy was defending, retreating Germans mined, burned, shelled and booby-trapped buildings, roads and bridges. Dazed civilians saw their homes and shops go up in flames, livestock killed their crops ground to dust, and loved ones buried alive beneath mountains of rubble. Captured bridges were kept intact to speed up the advance and prevent German armour crossing the bridges and attacking the landing beaches.

With regards to the destruction, six hundred of Normandy's towns and villages had to be completely rebuilt. Nearly one hundred and eighty seven thousand buildings were damaged, one hundred and thirty three thousand were completely demolished and over three hundred and fifty thousand people were left homeless. To this day many French debate whether the cost of liberation was too high a price to pay.

Churches and Church spires in particular that were used as outlook points and sniper posts by the Germans were mainly destroyed. Thousands of French civilians picked up the few possessions they could carry and tried to escape from the fighting. A fortunate few were able to ride bicycles or horse-drawn carts. But most people travelled on foot, pushing wheelbarrows or lugging their belongings on their backs.

Often they did not know where they were going or even where they would find their next meal. Some holed up in trenches, tunnels, caves and quarries. A few refugees even sought protection inside cemetery vaults and the padded cells of asylums. Others fled to nearby towns, villages or farms, only to be forced to flee again when the bombing and shelling caught up with them or when retreating Germans took over their shelters. Food was so scarce that refugees had to scramble to find enough food to live on. They butchered the animals killed by shells, salvaged food from wrecked stores and stole German supplies.

So great was the suffering that many civilians were too numb to celebrate when Allied troops finally arrived to liberate them.

At the time of the breakout the Allies enjoyed a considerable superiority in numbers of troops and armoured vehicles which helped overcome the natural advantages the terrain gave to the Germans. Although there were several well-known disputes among the Allied commanders, their main tactics and strategy were essentially agreed on. By contrast, the German commanders were bullied and their decisions interfered with by Hitler, who tried to control the battle from the distant Wolf's lair in East Prussia. With little knowledge of local conditions, he would intervene disastrously at tactical level. Hitler had never even visited the Atlantic Wall. All he did was boasting about it.

Field Marshals von Rundstedt and Rommel repeatedly asked Hitler for more discretion but were refused. Then von Rundstedt was removed from his command on 29th June after he bluntly told the Chiefs of Staff that the war was lost and to "Make peace, you idiots"!

The Battle of Normandy that lasted eleven weeks and was an overwhelming victory for the Allied forces. Paris was planned to be liberated on D+90 days, and despite many unexpected delays, it was achieved in a remarkable D+77 days. After the Normandy, break out, despite the bitter fighting that continued, the basic issue was no longer in doubt. The unconditional surrender of Nazi Germany was only a matter of time.

One of the great "what if's "in history is; "what might have happened if D Day had failed". However, in speculating there has to be a real chance that things could have turned out differently because of forces beyond human control; in this particular case, the weather.

An Allied repulse on the Normandy beaches would have been disastrous and the rest of the twentieth century would have been very different. The German generals would have been unlikely to risk their bomb assassination plot on what would have been a victorious Führer. Hitler could have redeployed east and bought time. The Soviets would have faced the full might of a rearmed and victorious

Wehrmacht, equipped by the vastly improved German Armaments industry. German arms production actually peaked in September 1944. New weapons such as V rockets, jet fighters, and snorkel U boats would have become available.

Today we take the D-Day success for granted. At the time, however, there was a real fear that the landings might fail and the Germans would hurl the invaders back into the sea. Churchill himself feared another first day of the Somme with its sixty thousand casualties. We even know that on the morning of 6th June 1944, Eisenhower secretly began to draft a signal beginning "The landings in Normandy have failed", just in case the invasion was the disaster it could so easily have been. Eisenhower might well have gone down as a defeated and disgraced supreme commander.

Throughout military history the cards fall where the wind blows them. The weather has on many occasions determined the result of battles; for example, The Spanish Armada and both Napoleon's and Hitler's Russian campaigns. But rarely have the whims of weather produced more far-reaching consequences than it did at D Day. Some parts of weather can be predicted in advance with a degree of certainty. But wind, waves, and cloud cover can scarcely be predicted much more than twenty-four hours in advance. Especially in an area of notoriously volatile weather such as the English Channel. Nowadays there are satellites and reporting stations such as weather men could never have imagined. But in those days it was more like guess work.

At the final weather conference scheduled for 04 00 on June 4th; Group Captain Stagg, whom Eisenhower described as a "dour but canny Scot", made the weather predictions, as he had been doing every day for a month. Stagg had bad news. The weather, that had been calm for the first few days of June suddenly deteriorated. June 5th would be overcast and stormy. Eisenhower decided to postpone the invasion for one day. In the early hours of June 5th, Stagg made what may be the most famous weather prediction in military history: He forecast that the storm would ease off later that day, and that by June 6th, the weather, although still poor, would be acceptable for landing conditions. The rain that was then pouring down would stop before daybreak. There would be thirty-six hours of more or less clear weather. Eisenhower made his decision and stated "Okay, let's go".

What if the storm had continued into June 6th? This is certainly within the realms of possibility. Eisenhower could have called the invasion back, although not easily. Had he done so, he would not only have given away the landing site, but June 19th, the next suitable date would witness the worst storm to hit Normandy in years. As previously stated in reference to the Mulberry's. If, on the other hand, had he gone ahead with the invasion and the weather had not cleared on June 6th, the consequences may have proved calamitous. The landing craft would have been tossed about like toy boats in a bathtub. Men trying to go ashore from any craft that made it to land would have been exhausted and incapable of fighting.

There would have been no air cover and no paratroop support, due to the air drops being scattered all over the place. No supporting bombing from the twin and four engine bombers. The Navy might have been able to fire its big guns, but because of the rolling seas, accuracy would have been limited. The German defenders, protected from the elements in their bunkers, would have delivered a deadly fire on the hapless Allied infantry. Eisenhower would have had no choice but to cancel the follow-up landings. He almost certainly would not have been able to withdraw many of the men from the initial landings. They would have been killed or captured, as had happened at Dieppe. At nightfall on June 6th, he would have issued his prepared-in-advance statement to the press: "The landings have failed".

Then what? Eisenhower would have certainly lost his job, and this was something he knew, which was why he had prepared his statement accepting full responsibility for the failure. There would be no sense bringing the entire high command down with him. But who could have taken his place?

Montgomery was unacceptable to the Americans. Bradley would have been seen to be tarred by the same brush of failure as Eisenhower. Perhaps Patton, who was preparing to take a field command after the beachhead was established. He would not have been implicated in the failed landing. George C. Marshall, the U.S.

Army chief of staff was a possible choice. He had originally hoped to lead the invasion but President Roosevelt felt that he was too valuable in Washington.

The Allied planners, meanwhile, would have been in despair. Despite failure, they still would have had an enormous force at their disposal of land, air, and naval forces. But it had taken two years to put the Normandy plan together and there was no alternative plan available. In retrospect, Normandy was the best choice; but the planners could not have tried there a second time. Where, then?

The Pas de Calais beaches were far better defended than those in Normandy, and Le Havre bristled with German guns. Reinforcing the South of France landings, Operation Dragoon in mid-August would have been the most appealing option, perhaps this was the only way to get the Allied forces into France. But such a diversion would have created immense logistical problems and leave the bulk of the Allied army much further from the Rhine and Berlin. Moreover, with his channel flank secure for the moment, Hitler would risk little in sending reinforcements south which was not the case when Operation Dragoon actually took place.

June 6th, 1944 was not just a decisive military event but also a political one that determined which ideological path Western Europe would follow in the next half century. The Churchill government could not have survived failure at D Day. Would the successor government negotiate with Hitler? Hardly possible and unthinkable by this stage! Hitler was convinced that if the Allied invasion was defeated, they would not attempt another. The Germans could then possibly fight the Soviets to a standstill.

In the United States, meanwhile, Roosevelt had a presidential election coming up. He would most likely have lost the election following failure at Normandy. The new Administration would probably prosecute the war in the Pacific more vigorously.

However, taking all of this into consideration, failure on D Day would not have spared Hitler the problems of a two-front war. The Allied forces still intact in Britain remained a threat. But he would have been free to transfer most of his army from France to the

Eastern front. Perhaps more important, he could have used the D Day failure to try and split the Grand Alliance of the United States, Great Britain and the Soviet Union. How difficult would it have been for Goebbels and the Nazi propaganda machine to convince Stalin that his western allies were ready to fight to the last Russian?

It is not inconceivable that Hitler and Stalin would have negotiated their way back to 1939, and reinstated the Nazi-Soviet pact. Alternatively, it is also possible that Stalin might have overrun Germany, then France, and the war in Europe would have ended with the Soviets in control of the continent. The Red Army would have been on the English Channel. It is hard to imagine a worse outcome.

The mounting Soviet threat would compel Britain and the United States to increase the severity of the bombing raids over Germany. A climax would have come late in the summer of 1945, with two atomic bombs dropped over German cities; thus creating a vacuum in a Central Europe devastated by atomic bombs.

In the Pacific in the summer of 1945, with the United States expending her atomic arsenal against Germany and Stalin free to transfer his armies from the German to the Japanese front, the Red Army would have invaded the northern Japanese home islands. In this scenario, Japan would have been spared the atomic bombing but subjected to a Communist dictatorship in the northern half of a divided country, as happened in Eastern Europe.

This was exactly what Stalin was planning and would have done if the Japanese had not surrendered. Had the Soviets invaded Japan, who knows when they ever would have left? We are fortunate that the Pacific war ended when it did. If the war had gone on longer, the entire East – West geopolitical situation might have changed. Harry Truman warned Stalin to keep away from the Japanese home islands, and the Soviet Dictator reluctantly agreed. This decision by Truman may have been as important as the one to drop the Atom bombs on Japan.

Taking all of this into consideration, it is most apparent that the consequences of a failure on D Day would have been a global catastrophe and what would have happened is a matter of conjecture. It is, however, obvious that one of the outcomes would *not* have been a Nazi Germany victory and almost surely, a Communist victory in Europe.

This would have meant no NATO and the possibility of a Communist Great Britain. That is a terrible prospect, but it might have happened if the Germans had thrown the Allies back into the sea on D Day. Thank goodness that "canny but dour Scotsman" got his weather forecast right.

~TWENTY~
MICHAEL WITTMANN

Michael Wittmann was the greatest exponent of the art of armoured warfare in military history. Resulting from his many campaigns, both on the Eastern front and in the West, Wittmann was awarded Nazi Germany's coveted decoration, the Knights Cross with Oak Leaves and Swords. The Knight's Cross decorations were instituted by Adolf Hitler to replace the Pour le Mérite also known as the "Blue Max", which was Imperial Germany's highest award in recognition of outstanding acts of gallantry.

Michael Wittmann was an officer in the Waffen-SS of which many books have been written. Many writers have been quick to tar those who served in front line units with the same brush as those who performed lesser tasks. However, more recently this subject has been examined far more objectively and in many cases, it has been justifiably concluded that premier Waffen-SS units could rightly claim that they were genuine élite military units. Moreover, their performance on the field of battle may be placed alongside and in many cases above their comrades in the regular Army. This is borne out by the fact that the Waffen SS received the vast majority of the Iron Crosses awarded during World War Two.

This chapter is not an attempt to whitewash the wartime record of the SS. Rather, it is dedicated to all the fighting personnel of the Second World War, and amongst them Wittmann and his crew, who, like the majority of their compatriots in the SS, had nothing to do with war crimes or atrocities.

Michael Wittmann was undoubtedly, the most successful tank commander in history. His ambush of the British 7th Armoured Division, the renowned Desert Rats elevated him to legendary status. His Tiger tank, almost single-handedly inflicted terrible losses and brought this entire Division to a standstill.

Michael Wittmann was born in 1914 on a Bavarian farm where he soon developed skills that were to assist him greatly in later life. In 1934, he, as did many of Germany's youth of this time, joined the rapidly expanding German Wehrmacht, which Hitler had released from the yoke of Versailles. Later in 1937 he joined the elite "Leibstandarte SS Adolf Hitler".

On the outbreak of war in 1939, he served in Poland as a driver and later in France as commander of a reconnaissance unit. After the fall of France, Wittmann fought in the Balkan campaign in 1941. Also in 1941, during the first day of Operation Barbarossa, Wittmann destroyed six Russian tanks and received the Iron Cross Second Class. Further actions in Russia saw him being awarded the Iron Cross First Class and a place in the officer training school. After graduating, Wittmann took command of the revolutionary Tiger tank, returned to Russia and in operation "Zitadelle," that proved to be the largest tank battle ever fought. He destroyed thirty tanks, mainly T-34's, plus twenty eight anti-tank guns and two artillery batteries.

In January 1944 Wittmann destroyed his eighty eighth enemy tank. Both he and his gunner Bobby Woll received the Knight's Cross for their exceptional skill, bravery, and outstanding gunnery tactics. That same month, he received the following telegram from Adolf Hitler: "In thankful appreciation of your heroic actions in the battle for the future of the Fatherland, I award you the Oak leaves to the Knight's Cross". Wittmann was decorated in person by the Führer at Wolfslair, East Prussia

In February 1944 Wittmann returned home to begin a new chapter in what was already an illustrious career. He was hailed as a national hero, and continually read about in the press. Photographs of this battle-hardened Panzer commander and stories of his exploits on the Eastern front were frequently published across full-page newspaper spreads. For the shy, unassuming Wittmann, the sudden thrust into the limelight was sometimes all too much, he continually stressed that he was only doing his job. Nevertheless, the propaganda machine was infatuated by this young SS officer. Wittmann finally decided to go along with the tide, to raise public morale. He got married in April 1944 and was seen everywhere, especially visiting tank factories, thanking workers for their great efforts in producing Tigers.

In May Wittmann was transferred to Normandy taking over command of 2nd Company, 12th SS Panzer Division. On D-Day, he moved to the invasion front, as part of the group under Rommel's command. On June 13th Wittmann was near Villers-Bocage, to the south of Caen where he engaged and destroyed a large contingent of the British 7th armoured Division. For this action he received Swords to his Knights Cross with Oak Leaves, being recommended by SS commander "Sepp" Dietrich. Adolph Hitler again in person decorated Wittmann and promoted him to the rank of SS Captain. He was at this stage the highest decorated German tank officer.

Following the D-Day landings, the Allied failure to take Caen did massive damage to Montgomery's reputation, especially with the Americans, who viewed him as a losing general. Caen dictated British tactics; remember it had been planned to be taken on D Day. But by 13th June, a full week after the beach landings, Allied formations including the famous 7th Armoured Division only reached the vicinity of the city. The Allies had massive air superiority and the flanks of Wittmann's Division were dangerously exposed - setting up the possibility of being completely encircled. But, right in the path of the British was the village of Villers-Bocage.

The German tank commanders concluded that this important road junction would have to be secured. Wittmann was ordered by his battalion commander to halt the British armour before it could outflank and envelope the Germans. Nobody could have predicted the events that were to follow.
Wittmann set out towards Villers-Bocage, moving cautiously to avoid

being spotted from the air. He observed the British column passing him at a distance of two hundred meters. The British were completely unaware of the presence of SS armoured units in the area. The officer in command had requested time to carry out a proper reconnaissance. Orders were, however, to push on regardless. This decision was to have dire consequences. The British advanced accordingly, but, to clear the traffic on the roads, the armoured column had to move comparatively closed up. Due to the traffic jam up, the British commanders decided upon a staff discussion and briefing. Some British troops even began to brew up, thinking the area was safe. And it was this that gave Wittmann his opportunity.

Wittmann immediately saw the column's vulnerability and decided to go into action. The road was packed with British vehicles, including Sherman, Stuart and Cromwell tanks, plus Bren-gun carriers and half-tracks. British troops were lounging around, smoking or drinking tea. Wittmann's gunner, Bobby Woll, remarked: 'They're acting as if they've won the war already'. To which Wittmann replied: 'We're going to prove them wrong'.

Leaving three Tigers and a Panzer Mark IV to protect his flank, Wittmann thundered towards the British along a sunken lane into Villers-Bocage, where he had a field day with his 88 mm gun. The heavy Tiger poured shell after shell into the mass of machines. His first shot destroyed one of the half-tracks, thus blocking the road; and then at his own convenience he destroyed the remainder of the half-tracks and the four Cromwells of Regimental Head Quarters.

Half-tracks, carriers and tanks were smashed, and then with a final burst of speed the fifty five ton steel monster crashed through the junction and swung onto the roadway to begin its descent upon the vehicles lined up outside the village and along the narrow high street. Escape for the tanks, carriers and half-tracks was impossible; the road was obscured by flames and smoke from the burning vehicles whose crews could only seek what shelter they could.

The British position was untenable. In addition to the burning tanks and vehicles; the road was blocked by the German Tiger that commanded all approaches. Wittmann's Tiger attacked and destroyed three further tanks; a Sherman Firefly and two Cromwells.

Proceeding without pause he attacked the lightly armored vehicles of The Rifle Brigade. During this engagement, he destroyed nine half-tracks, six troop carriers, and two 6-pounder anti-tank guns. He then destroyed three Stuart light tanks and another half-track. Wittmann continued his advance, destroying three further Cromwells. He then engaged and destroyed two Sherman's before knocking out another scout car and half-track.

As Wittmann arrived at the town square, he was opposed by a Sherman whose 17-pounder gun was the only Allied main tank capable of taking on a Tiger. Four shells were fired at Wittmann's Tiger, one hit the hull, and he returned fire and knocked out the Sherman. A Cromwell tank that had not been destroyed, confronted him, firing two 75mm shells that bounced off the thick German armour, failing to harm the Tiger. Wittmann put the Cromwell out of action with one shot.

The petrol driven Sherman's burned so easily to earn the nickname "Tommy Cooker", and the unfortunate crews knew that more often than not, when the Sherman 'brewed up' they only had a fifty-fifty chance of baling out. Wittmann's Tiger knocked out three more British tanks and then started to withdraw. But, as he left Villers-Bocage, his tank's left track was hit by an anti-tank shell and immobilised. He then engaged targets within range before abandoning the tank and together with his crew made their way on foot to German lines. Almost single-handedly, Wittmann had destroyed the British advance and forced the 7th Armoured Division to retreat. The British withdrew from Villers-Bocage and set up a defensive position to the north-west of the town, leaving it to the Germans, who occupied it for the next two months.

Wittmann's action derailed a major flanking movement that posed a significant threat to German positions. He had taken action on his own initiative against a significantly larger force, that had total air control, and still within range of massed naval guns. The German attack at Villers Bocage had effectively cut the British line in two. During this short engagement, Wittmann had destroyed seven Sherman's, twenty Cromwell's, three Stuarts, fourteen half-tracks, sixteen Bren Carriers and two - 6 pounder anti-tank guns. The British drive on Caen was stopped cold by Wittmann's attack. The engagement at Villers-Bocage marked the end of a British attempt to bypass the key town of Caen, only sixteen kilometers inland from the Anglo-Canadian landing beaches.

In considering Michael Wittmann's success, it must be remembered that the British made a number of basic errors that are highlighted as follows; firstly, they were moving across the enemy's front without due regard to security and without a clear idea of the presence of the enemy that had heavy forces. This was a highly dangerous position to be in. Secondly, no proper reconnaissance was done, although to the commander's credit, he had made such a request but was denied; and thirdly, seemingly oblivious to security, they stopped for a briefing and discussion in the middle of a battlefield. This was careless to say the least.

These errors could only be attributed to either complacency or inexperience - probably both. But there were mitigating factors. Whilst the 7th Armoured Division itself was a veteran unit, having fought in North Africa and Italy, not all regiments within the Division were veterans. Some were very green indeed, and had not learned the hard lessons of battle.

Returning to Michael Wittmann; blowing up so many tanks and other vehicles around Villers-Bocage was certainly impressive. But we've already seen that this could be attributed to the British lack of preparedness and also carelessness. If the British defences had been well organised, Wittmann's career might well have ended right then and there instead of two months later.

The main reason why Wittmann did so well in that assault is that he was free to move on his own choosing. The Germans generally had a higher level of combat experience in Normandy and the terrain vastly favoured the defender and somewhat nullified the benefits of Allied air superiority. Also it should be remembered that the bulk of the German armoured force in Normandy was directed against the British/Canadians rather than against the Americans.

What is without doubt is that Wittmann's Tiger assault accomplished the goal of the whole German strategy: to prevent a British flanking move. Taking into consideration that you do not become a tank ace by going on Rambo missions; you become an ace by being able to exploit an enemy failure, and also being able to prevent your own losses.

After Villers-Bocage, Wittmann was offered the position of a training instructor. He refused, remained in Normandy, got a new Tiger #007. He then participated in the battle for the Falaise Pocket. This proved to be his final action. On 8th August 1944, his Tiger was destroyed and the entire crew killed.

Wittmann's company was situated at a small village to the east of Falaise. His opposition was none other than an armoured column of the Black Watch of the 51st Highland Division under the command of Lieutenant-Colonel Douglas Forster. German armour set off northwards, crossing the open fields at top speed with the 'Hitler Jugend' panzer-grenadiers following behind. The tanks were met with Allied artillery fire, but Wittmann pushed on even faster and got through unscathed.

At one thousand eight hundred metres, using their superior range the Tigers began to shoot up the advancing Sherman's who were receiving horrific casualties. Lieutenant-Colonel Forster was badly wounded by friendly aerial fire, as occurred frequently in Normandy, and had to be evacuated. When Wittmann's Tiger's were at a range of one thousand two hundred meters, Captain Boardman, the British second in command, ordered his unseen Sherman's to hold fire to enable him to move to a better position in order to control the firing. Boardman had only four tanks with armament capable of penetrating a Tiger's armour at close range.

Meanwhile Wittmann's three panzers still continued on their course in line ahead closing the range to one thousand meters. In doing so they exposed their right flanks, presenting ideal targets. Wittmann had not been informed that the British had 17-pounder Sherman's in the area. At eight hundred meters Captain Boardman gave the order to open fire. The Sherman's engaged the rear Tiger, firing two shots which knocked it out.

The second Tiger traversed right and fired three shots. All three shells missed as the Sherman's moved to a new firing position. One shot was fired at the second Tiger and it exploded. By this time Wittmann's leading third Tiger came under heavy fire from the rest of the Sherman's to settle the matter. Thus, in less than twenty minutes, the British had finally dispatched the most distinguished of all German ace tank commanders. SS-Captain Michael Wittmann had fought and lost his last battle.

According to the original War Diary of 22 Armoured Brigade, on 8th August at from 12:20 hours, three Tigers were destroyed without any British losses. It was claimed that Wittmann's Tiger was destroyed by Sherman's armed with 17 pounder guns, capable of penetrating a Tiger's armour at a range of eight hundred meters. How accurate is this version? Several conflicting reports have been produced about whom and what exactly delivered the final blow.

Immediate examination of Tiger tank #007 showed there were also signs that Wittmann's Tiger could have been destroyed by a British Typhoon fighter-bomber. But, whether the Typhoon damage was inflicted "postmortem" or whether the tank was still operational and Wittmann alive are still unclear.

Several sources claimed to have ambushed and destroyed Wittmann's Tiger, including 1st Polish Armoured Division, 4th Canadian Armoured Division, and 33rd British Armoured Brigade. Some claims were made by units which were not even present in the area at the time. The official version at the end of the war stated that Wittmann's Tiger had been destroyed by air attack. A picture thought to be Wittmann's Tiger without its turret was presented. But this was in fact a picture of Alfred Günther's Tiger that had been destroyed by an air attack.

In 1945, Wittmann's Tiger was located and examined. It was observed that it had not been penetrated by any tank shells fired during the fighting. The only damage was a gaping hole in the rear, and stated that the impact came from the air. A Typhoon rocket hit the Tiger's rear deck of 25mm thick armour, penetrated the air intakes, exploded and ignited the stored ammunition. The ammunition explosion instantly killed the entire crew and blew off the turret. Therefore, it was concluded that Wittmann's Tiger was destroyed by a rocket fired from a Royal Air Force Hawker

"Typhoon".

Typhoons were armed with High-explosive rockets and took a heavy toll of German tanks during the Normandy battles. They actually destroyed one hundred and thirty five German tanks and among those claimed was Wittmann's Tiger. Unfortunately no photographs exist of Wittmann with his crew at the August 8th battle.

Captain Michael Wittmann and his crew have been credited with destroying one hundred and thirty eight tanks and one hundred and thirty two anti-tank guns when they were killed. This record has never been equaled in warfare. Wittmann was by far the most successful tank commander of the Second World War, indeed, of all time. He had a combination of acute observation, strategic and tactical intuition, and was able to create surprise assaults that caused havoc, throwing his enemies totally off balance.

Coupled with this, his man management skills were quite exceptional. He regarded his crew members as individuals of an elite combat team, not as parts of the tank. He was highly admired by his comrades and superiors. Especially his gunner "Bobby" Woll who could interpret his commander's decisions resulting in a highly coordinated and extremely efficient fighting team.

Regular gunner Bobby Woll, holder of Iron Cross 1st and 2nd Class and Knight's Cross, was not part of the crew at Wittmann's last battle. He had received command of his own Tiger and took part in the Normandy battles, where he was wounded during an air attack and remained in hospital until March 1945. He returned to active service and took part in the final battles on the Western Front. He survived the war, became an electrician and died in 1996.

Wittmann and his crew are buried in the German Military Cemetery of "De La Cambe" in Normandy.

SS-General "Sepp" Dietrich said after Wittmann's death.
"He was a fighter in every way, he lived and breathed action".

Michael Wittmann fought to the bitter end for his Fatherland; his place in the annals of military history is thoroughly deserved.

TWENTY ONE
MANSTEIN - SUPREME STRATEGIST; HITLER - SUPREME COMMANDER

Clausewitz, in his writings on the "Art of War" said everything in war is simple, but the simplest thing is incredibly difficult. Consider the basic relationship between politics and war. Clausewitz stated that war is an extension of politics by different means. Politicians establish the objectives; the military seek to attain them. Therefore, politics are thrust upon the military, to do their duty and await their fate. Nothing could be simpler.

Except perhaps not in the case of Adolf Hitler, who gained remarkable political triumphs. He occupied the Rhineland, then annexed Austria, and Czechoslovakia; thereby expanding German territory without resorting to force. When he used military means in Poland to obtain his political aims, he demonstrated the close connection between politics and war. But, his invasion of Poland precipitated World War Two and from then on, his direction of the war became increasingly military and less political.

The result was that towards the end, the fighting degenerated into senseless destruction for the sake of merely continuing the struggle. He no longer had political objectives only total war. The consequences of which was the deterioration of the military and ultimate defeat. Highlighting this scenario; in the summer of 1940,

Germany overran Denmark, Norway, and the Low Countries. On 14th June the Germans entered Paris, compelling the French to seek an armistice. Hitler's Germany was master of Western Europe and only awaited the British to accept terms.

Erich von Manstein's planned Sichelschnitt or Sickle cut had succeeded beyond all expectations. The scythe had taken its harvest. France was beaten and Germany rejoiced. Manstein's plan had eliminated the French and British armies in continental Europe and achieved what Imperial Germany had failed to do twenty-five years before.

Manstein, perceptively, posed the question. What next? It was now obvious to him that Hitler had no long-range plans, and could neither conclude peace nor invade Britain. The question of crossing twenty miles of sea had not arisen. When it did arise in June 1940 there was no answer. Herein lay the major flaw, not in what *Sickle cut* had done, but in what Hitler as supreme commander had failed to do. Especially when combined with the fatal mistake of halting the panzer's outside Dunkirk; thus allowing the British to evacuate their troops back across the Channel.

What was Sickle cut? – Later to be referred to as the 'Manstein Plan'.

Von Manstein's plan of the 1940 German offensive required a thrust through the Ardennes, the line of least expectation. It achieved the decisive break through on the Western Front, and led to the fall of France. At this time Manstein was Chief of Staff to Rundstedt's Army Group. His tenacious arguments to change the modified First World War Schlieffen plan had become irritating to his superiors.

Consequently he was pushed out of the way to command a reserve corps of infantry. However, when the original German attack plans fell into Allied hands, Hitler flew into a rage and demanded a revision. Manstein again forwarded his proposals. On learning of Manstein's plan for the offensive, Hitler immediately grasped the idea. The new plan was adopted and succeeded spectacularly to trap the Allies on the Channel coast.

But, the next step had not been thought through. This lack of foresight was the first evidence of the manner in which Hitler would conduct his war. Decisions would be taken on the spur of the moment, with little proper planning or consideration for the implications.

When the problem of Britain became apparent in June 1940 it was already too late. The German Navy was ill prepared. Göring had insufficient aircraft of the type required. The Luftwaffe had been created primarily as an Army support-arm, not as a strategic bombing force. Even had the Luftwaffe secured the sky above the English Channel the Navy could only muster sufficient forces to protect a narrow corridor between the two coasts. Neither were there enough craft, certainly none designed specifically for this purpose. An invasion in the summer of 1940 was never a practical possibility.

So what was to be done about Britain? Before the war Hitler assumed that Britain would remain neutral and he made virtually no preparations against her. His offer of peace produced only a growl of defiance. Clearly a tiger had been taken by the tail. Hitler saw the Channel as Britain's salvation, not his own failure to plan for a protracted war. Hitler, therefore, decided that if *Blitzkrieg* was inapplicable to Britain, then it was applicable elsewhere on the European continent.

Victory in France had been so swift that Hitler was now convinced he was infallible. He did not take into consideration that France had been an ideal environment for panzer warfare. The distances were short, the roads and weather good and casualties low. Everything had gone according to plan. But not many countries had the road network or the political vulnerability that existed in France in 1940. Nevertheless Hitler decided that the Soviets were for political and economic reasons his next target. It is impossible to judge whether or not the decision to turn on the Soviet Union was unavoidable for military reasons. Certainly Stalin had sanctioned the buildup of troops on the Nazi- Soviet border.

On the commencement of the invasion of Russia in 1941 Manstein was given command of an armoured corps. He made one of the quickest and deepest thrusts of the opening stage, nearly two hundred miles in four days. Promoted to command the Eleventh

Army in the south, he forced an entry into the Crimean peninsula, and in the summer of 1942, captured the famous fortress of Sevastopol, the key naval base on the Black Sea. Manstein's appointment as an army commander brought him for the first time under Hitler's direct orders.

He was now to experience how Hitler fulfilled the responsibilities of supreme military commander combined with Head of State. Previously, during the campaign in Poland Manstein had been unaware of any interference by Hitler in the military leadership. Hitler had listened sympathetically to military interpretations of the situation and made no attempt to intervene. He undoubtedly had an eye for operational openings, as had been shown by the way he opted for Sickle cut in the west. In addition, Hitler possessed an astoundingly retentive memory and an imagination that made him quick to grasp all technical matters and problems of armaments.

He was amazingly familiar with the effect of the very latest enemy weapons and could reel off whole columns of figures on both German and the enemy's war production. Indeed, this was his favourite way of side-tracking any topic that was not to his liking. He prided himself with the production figures of the German armaments industry, which he had boosted to an amazing extent; preferring to overlook the fact that the enemy's armaments figures were immensely higher.

What he failed to appreciate was the amount of training and skill required to render a new weapon fully effective. It did not worry him whether a weapon had even been tested under combat conditions. There can be no question that his insight and unusual energy were responsible for many achievements in the sphere of armaments. His belief in his own superiority in this respect ultimately had disastrous consequences. His interference prevented the smooth and timely development of the Luftwaffe, and he hampered the development of rocket propulsion and atomic weapons.

Hitler lacked military ability based on experience; something for which his 'intuition' was no substitute. He failed to understand that the objectives of an operation must be in direct proportion to the time and forces needed to carry it out. He did not realise that any long-range offensive operation calls for a steady build-up of reserves over and above those committed in the original assault. All this was brought out with striking clarity in the planning and execution of the 1942 summer offensive in Russia.

Another example was the fantastic idea he disclosed to Manstein in the autumn 1942 of driving through the Caucasus to the Middle East and India with a motorized army group. What he lacked was experience in strategy and grand tactics. His active mind seized on almost any aim that caught his fancy, causing him to fritter away Germany's strength by taking on several objectives simultaneously. The rule that *one can never be too strong at the crucial spot,* to achieve a decisive aim, was something he never really grasped.

As a result, in the offensives of 1942, and 1943 he could not bring himself to stake everything on success. Neither was he able or willing to see what action would be necessary to compensate for any unfavourable events that may occur. Hitler became obsessed with the symbolism of capturing Stalingrad named after his arch foe, regardless of its strategic value. When the magnitude of the disaster became obvious, Manstein was sent to conduct the efforts to relieve Paulus's Sixth Army, trapped that winter at Stalingrad.

The effort failed because Hitler refused to agree with Manstein's insistence that Paulus should break out westward and meet the relieving forces. The relief column actually got to within fifty kilometers, but Paulus did not break out to meet it, unless he received direct orders from Hitler. Approval could not be obtained. Göring had assured Hitler that his Luftwaffe would supply Stalingrad.

Manstein, a brilliant strategist, now perceived the obvious. He realised that the chances of bringing out 6th army intact was hopeless. He was forced to abandon the rescue mission when a Russian counter stroke threatened him with encirclement. In fortress Stalingrad the conditions of the stricken army rapidly deteriorated. The exhausted troops were reduced to scavenging and eating horseflesh. Paulus bitterly complained that the Luftwaffe had left him in the lurch.

Temperatures had plunged to -30 degrees C, and due to Göring's incompetence, hardly any winter clothing had arrived. The men were suffering from frostbite, dysentery and typhus with little hope of survival. The airlift was a chronic failure. Hitler must be held accountable for placing the Third Reich in a hopeless position, and Stalingrad was his death blow.

Manstein informed Hitler that due to the Stalingrad disaster the position in the Caucasus had become untenable. The German forces had to be withdrawn immediately from the Caucasus. The Wehrmacht's situation was critical. Yet, it had to hold on in Stalingrad, for surrender would unleash an even greater catastrophe. Hitler eventually relented, and gave the necessary order to withdraw from the Caucasus. Manstein, by skillful military manoeuvering, managed to evacuate the endangered German armies which Hitler had recklessly and fatally thrust south into the Caucasus. Total disintegration on the eastern front was thus avoided. Manstein, could not save the trapped 6th army, but he did rescue the armies in the Caucasus from a similar fate.

By a brilliant flank counterstroke, Manstein recaptured Kharkov and rolled back the Russians in confusion. That counterstroke was the most brilliant operational performance of Manstein's career, and one of the most masterly in military history.

The last great German offensive in the East, 'Operation Citadel', was launched in July 1943 against the Kursk salient. Hitler again interfered and several postponements led to Manstein's despair of success. Manstein's Southern Army Group formed the right pincer. It achieved a considerable measure of success, but the effect was nullified by the failure of the left pincer.

Moreover, at this crucial moment Anglo-American forces landed in Sicily; concerned that his longtime friend and ally Mussolini might collapse. Hitler diverted several divisions to Italy, thus halting the German attack at Kursk. The Russians then launched their own offensive on a larger scale, and with growing strength. From that time onwards the Germans were thrown on the defensive. Hitler no longer aimed to win the war. He aimed to keep the war going until his enemies wearied and agreed to a negotiated peace.

Manstein was called on repeatedly to conduct a fighting withdrawal in the face of much superior forces. He showed great skills in checking successive Soviet thrusts and imposed delays on the westward advance of the Russian armies. He constantly looked for opportunities to deliver counter attacks. But when he urged that a longer step back should be made, that is, a *strategic* withdrawal, to facilitate the development of the full recoil spring effect of a counter-offensive against an overstretched enemy advance. Hitler would not hear of it.

His unwillingness to sanction any withdrawal forfeited any chance of stabilising the front, Manstein maintained the Old Prussian tradition of speaking frankly, and expressed his criticism forcibly. Hitler repeatedly clashed with Manstein's defensive strategy. In the end it became more than Hitler could stand; particularly as the course of events continued to confirm Manstein's warnings. So in March 1944 Hitler dismissed Manstein; thus ended the active career of a man who combined modern ideas of mobility with a classical sense of manoeuver.

Hitler's strategic aims were to a very great extent conditioned by political considerations and the needs of the German war economy. Political and economic goals are undoubtedly of great importance. However, what Hitler overlooked was that the achievement and, most importantly of all, the retention of objectives requires the defeat of the enemy's armed forces. Otherwise the attainment of territorial aims and their long-term retention is a sheer impossibility.

The strategic aim of any war is to smash the military power of the enemy before the way is open to the realisation of political and economic aims. This brings the factor which probably did more than anything else to determine the character of Hitler's leadership; his overestimation of the power of the will as one of the essential prerequisites of victory. Admittedly, many a battle has been lost and many a success thrown away because the supreme leader's will failed at the critical moment.

However, the will for victory which gives a commander the strength to see a grave crisis through is something very different from Hitler's will. He became impervious to reason and thought his own will could operate beyond the limits of hard reality. The essential elements of the 'appreciation' of a situation on which every military commander's decision must be based were virtually eliminated. With that Hitler discarded reality together with the enemy's resources and possible intentions.

The same man who, after his decisive successes in politics up to 1938, actually recoiled from risks in the military field during wartime. In the Russian campaign, Hitler's fear of risk manifested itself in his refusal to accept that elasticity of operations; which, in the conditions from 1943 onwards, could only be achieved by a voluntary, surrender of conquered territory.

There was also his fear to transfer troops from secondary fronts in favour of the spot where the main decision would fall; the principle reason why Hitler evaded these risks in the military field was the fear, common to all dictators, that his prestige would be shaken by any perceived setbacks. Hitler had an intense dislike of giving up anything on which he had once conquered. Whenever he was confronted with a decision which he did not like taking but could not ultimately evade, Hitler would procrastinate as long as he possibly could. This happened every time it was urgently necessary to commit forces to battle in time to forestall an operational success by the enemy.

The General Staff had to struggle and battle in vain with Hitler to get forces released from· less-threatened sectors of the front to be sent to a crisis spot. In most cases he would give too small a number of troops when it was already too late. Hitler always expected things to go his way in the end, thereby enabling him to avoid decisions which were repugnant to him; his inflated belief in his own will-power, and an aversion to accepting any risk when its success could not be guaranteed in advance influenced Hitler's military leadership. Obstinate defence of every foot of ground gradually became his style of leadership.

Hitler's reaction when the crisis occurred in front of Moscow was to adopt Stalin's precept of hanging on doggedly to every position. It was a policy that had brought the Soviet leaders so close to defeat in 1941 that they finally relinquished it when the Germans launched their 1942 offensive.

Hitler was convinced that his ban on any voluntary withdrawal had saved the Germans from the fate of Napoleon's Grand Army in 1812. When, therefore, a fresh crisis arose in autumn 1942 after the German offensive had become bogged down at Stalingrad, Hitler again thought success lay in clinging on at all costs to what he already possessed.

It is generally recognized that defence is the stronger of the two forms of fighting. This is only true, however, when the defence is so strong that the attacker bleeds to death when assaulting the defender's positions. Such a thing was out of the question on the Eastern Front, where the number of German divisions available was never sufficient for so strong a defence to be organized. The enemy, being many times stronger, was always able, by massing his forces at points of his own choice, to break through fronts that were far too widely extended. As a result, large numbers of German forces were unable to avoid encirclement. Only in mobile operations could the German defenders repel the advancing enemy.

The effects of Hitler's 'hanging on at all costs' may be found deep down in his own personality. He was a man who saw fighting only in terms of the utmost brutality; substituting the *art* of war with brute force and will-power. Hitler was certainly quite clearly informed of conditions at the front through the reports he received. In addition, he frequently interviewed officers who had just returned from the front-line areas. Thus he was aware of the continuous overstrain they had had to endure.

This could still have been counterbalanced, if he had been prepared to take advice from, and place genuine confidence in, an experienced Chief of the General Staff. This would have compensated for his lack of training and experience in the military sphere, particularly as regards strategy and grand tactics. By utilizing the skill of his Chief-of-Staff, quite an efficient military leadership might have emerged. But this was precisely what Hitler would not accept. He considered the power of his will to be in every way decisive.

As had his political successes, and, indeed, the military victories early in the war, which he regarded as his own personal achievement. This caused him to lose all sense of proportion in assessing his own capabilities. To him the acceptance of advice from a Chief-of-Staff would not have meant supplementing his own will but submitting it to that of another. Added to this was a mistrust of the military leaders, whose code and way of thinking were alien to him. Thus he was not prepared to have a really responsible military adviser alongside himself.

He wanted to be another Napoleon, who had only tolerated men under him who would obediently carry out his will. Hitler had neither Napoleon's military training nor his military genius.

Hitler thought he could see things much better from behind his desk than did the commanders at the front. He ignored the fact that much of what was marked on his situation maps was out of date. From that distance, moreover, he could not possibly judge what the proper and necessary actions to take on the spot were. In any discussion of operational intentions one is almost always dealing with a matter whose outcome nobody can predict with absolute certainty. Nothing is certain in war.

The more he came to regard the principle of 'holding on at all costs' policy, the less prepared was he to issue long-term directives. His mistrust of his subordinate commanders prevented him from giving them, in the form of long-term directives, freedom of action, which they might put to a use that was not to his liking. In the long run even an army group could not function without directives from the Supreme Command. Whenever Hitler perceived that he was not making any impression with his opinions on strategy, he immediately produced something from the political or economic sphere. Which no front-line commander could compete, his arguments here were generally irrefutable.

Naturally no military relationship could develop between this dictator, who thought only of his political aspirations and lived in a belief in his 'mission.' Manstein made no less than three attempts to persuade Hitler to accept some modification of the Supreme Command. He was fully aware that Hitler would never be prepared to relinquish the supreme command officially. As a dictator he could not possibly have done so without suffering what for him would have been an intolerable loss of prestige.

Everything depended, therefore, on persuading Hitler, whilst nominally retaining the position of Supreme Commander, to leave the conduct of military operations in all theatres of war to one responsible Chief-of-Staff and to appoint a special Commander-in-Chief for the Eastern theatre. These attempts proved unavailing. Hitler knew full well that Manstein was the very man many people in the army would like to see as Commander-in-Chief in the east.

With regards to the changing the leadership of the Reich by violent means, as on 20th July 1944, Manstein would not contemplate a *coup d'etat* in wartime because he considered it would have led to an immediate collapse of the front and probably to chaos inside Germany. Apart from this, there was always the question of the military oath and the committing of murder for political motives. Stating 'No senior military commander can expect his soldiers to lay down their lives for victory and then precipitate defeat by his own hand'. In any case, it was already clear by that time that not even a *coup d'etat* would make any difference to the Allied demand for unconditional surrender.

Manstein was trusted by his staff and by his commanders as a man who thought carefully. He only took risks when he felt them justified and usually managed to out think the enemy. His peers regarded with near unanimity that Manstein was their finest brain and best commander and this has been echoed by military commentators from the Allied countries. They called Manstein 'the most brilliant and gifted strategist of all German generals', always full of new and often brilliant ideas, an organizer of genius, he was always in the front rank where the interests of the Army were at stake'.

'The ablest of all the German generals was the verdict of most of them on Field Marshal Erich von Manstein'.

Manstein's favourite tactic was to allow a Russian penetration in a particular area and then encircle the attackers when they had been lulled into complacency. This successful ruse showed that he did not think merely in terms of fixed lines. When taxed with losing

Kharkov, Manstein said: "I'd rather lose a city than an Army".

Hitler's own opinion of Manstein, given to Jodl in 1944, was this: 'In my eyes Manstein has a tremendous talent for operations, there's no doubt he is certainly one of our most competent officers. He believed that it was still possible to force a stalemate on the Eastern front, but his plans involved withdrawals that Hitler would not countenance. Manstein constantly had to fight with under-strength formations.

Returning to 1940, Hitler saw Sickle cut as a personal triumph, over his own General Staff, considering them to be lacking in imagination and resolve. The vindication of victory removed any remaining inclination to listen to his professional advisers. This made him more inclined to assume the right to interfere in the military conduct of war,

In the summer of 1940 Hitler held sway over the continent of Europe. Britain nursed her pride and prepared for the Luftwaffe's onslaught; on the other side of Europe the Soviet Union held its breath and rushed more troops into Poland. Manstein stated that Germany's power had never been higher. Up to 1940, the German military was united as never before. The Führer and generals, all worked together. The result was triumph. It was the last time they all worked in harmony. The consequence, not surprisingly, would be tragedy for the Third Reich.

In June 1941, Hitler totally miscalculated both his resources and the immensity of the task when he invaded the Soviet Union. Unable to determine which political and economic targets to pursue, he fragmented his military forces and in the end lost all. Hitler's policies of supreme command, led to growing disenchantment among the officer corps. The dilemma for all German professional soldiers like Manstein was that of trying to serve their country while disapproving the Führer's aims and methods.

True to their tradition of blind obedience, most of them concentrated on the military role and ignored politics, whilst deploring the lack of political direction. They witnessed the strategic opportunities that were missed and the bright prospects that were ruined.

After the war von Manstein was tried as a war criminal. He was charged with seventeen offences, including responsibility for the execution of Jews in Poland and Russia; plus the deportation of civilian populations, the execution of Hitler's Commissar Order and unjustified reprisals against, partisans. It emerged after the trial that most military commentators thought Manstein had conducted himself very well as a commander.

Far from instigating any policies of brutality he had done his best to make life more tolerable for the Russians in his occupied areas. It would have been impossible for Manstein to personally supervise the whole of the vast areas under his control and undoubtedly atrocities did occur, but Manstein did not countenance them. Nevertheless, the court found him guilty on two charges and he was sentenced to eighteen years imprisonment. This sentence provoked opposition from several distinguished quarters.

One letter to The Times newspaper ended:
"I have studied the records of warfare long enough to realize how few men who have commanded armies in a hard struggle could have come through such a searching examination of their deeds and words as well as Von Manstein did".

The original sentence was reduced greatly and Von Manstein was released in 1952. Following his release he wrote his memoirs before he died in Bavaria in 1973.

Erich Von Manstein was born in Berlin in 1887. In 1906 he joined an infantry Regiment and by the beginning of WWI he had reached the rank of lieutenant. In 1915 he attended the War Academy and following his graduation he returned to his Regiment where he fulfilled a number of Staff positions until the war's end. He gradually climbed through the ranks, and by 1937 had reached the rank of Colonel and was appointed to the General Staff.

He was promoted to Major General and appointed as Deputy Chief of the General Staff. In 1938 he was given command of the 18th Infantry Division. In 1939 prior to the invasion of Poland, he was appointed as the Chief of Staff for Army Group A, under Von Rundstedt. Following the conclusion of operations in France he was promoted to General and awarded the Knight's Cross.

After the war, German military commanders would claim that Hitler's grandiose plan was never really feasible and totally beyond the capacity of the Wehrmacht. Manstein, dismissed by Hitler in March 1944, sat out the rest of the war, watching, no doubt with dismay, the unnecessary prolongation of a conflict that had already been decided.

Although he served an evil and brutal regime, he was patriotically motivated to fight for his country. He maintained the highest personal standards of character and became the officer most widely respected and admired by his colleagues. The overall verdict among the German generals interrogated in 1945 was that Field-Marshal von Manstein had proved the ablest commander in their Army. He was the man they had most desired to become its Commander-in-Chief. It is very clear that he had a superb sense and mastery in the conduct of grand strategy.

Manstein was later called on by West German Chancellor Konrad Adenauer in 1956 to act as a senior advisor in the effort to help build the new German army and thereby enable Germany's entry into NATO.

~TWENTY TWO~
EUROPE'S SOFT UNDERBELLY

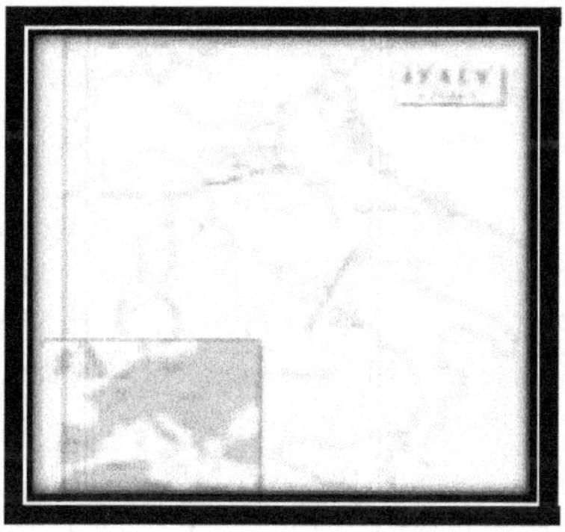

When Italy entered the war on June 10th 1940 the bombastic ravings of Mussolini could not disguise the country's general apathy and the widespread lack of support for his ambitions. This changed initially when there appeared to be hope of a swift Axis victory and considerable spoils to be gained. In the 'New Europe' Italy would be rewarded with rich pickings. But when successive Allied blows brought the war nearer to the Italian homeland, the people turned on and ousted Mussolini. However, this could not prevent Italy becoming a battlefield as German and Allied troops poured in to fill the vacuum.

Returning to the early days after Italy entered the war, Hitler did not want a conflict on his southern flank, and therefore, restrained his Axis partner on several occasions from putting into operation plans for an Italian invasion of Greece. Yet on the morning of October 28th 1940, Italian forces based in Albania crossed the frontier into Greece. To understand why the Italian dictator ignored Hitler's wishes, it is necessary to evaluate the situation from Mussolini's viewpoint.

He was a man obsessed with ideas of personal greatness and genius. To satisfy this he proclaimed the birth of a 'New Roman Empire' which was to take shape under his direction, and would recreate the ancient glories of Rome.

Italy had already overrun two weaker countries, Abyssinia in 1936 and Albania in March 1939. Mussolini recognised that he did not possess the raw materials to make war on the grand scale. This, however, did not prevent him from making it clear that Italy was ready to fight by Germany's side. Hitler made it clear that Germany did not need Italy's help. He wanted an asset and not a liability for an Axis partner.

When Mussolini sent his armies into Greece, far from winning acclaim for Italian arms, the disasters brought down the scorn not only of the Allies but also of Hitler. When the Italians pounced on Greece without Hitler's knowledge, the defenders replied with counter attacks that drove the aggressors back over the frontier into Albania. Eventually the Germans were drawn in, resulting in the delay of attacking Russia by a critical six weeks.

This would not be the last time that Hitler was again called upon to rescue his rash partner when Rommel and his Afrika Corps came to his aid in North Africa. Prior to this a mainly South African force routed the Italians in Somalia and Ethiopia.

But it was not just on land that the Italians suffered grievously. The Royal Navy completely outclassed the Italian navy on more than one occasion, especially at Taranto.

The decision to invade southern Italy stemmed from the Casablanca Conference in January 1943, while the Tunisian campaign was still in progress. The purpose was to secure Mediterranean lines of communication, divert German strength from the Russian front, and force Italy out of the war.

The British and Americans disagreed on their strategy towards Italy. The British plan was to keep the Germans tied down in the Mediterranean for as long as possible before the Allies made their main attack across the Channel. They therefore proposed the invasion of Sicily as a stepping-stone to the Italian mainland, hoping to draw Hitler into a major campaign in the Mediterranean. The assault on Italy might also cause the fall of Mussolini. Moreover, bases in Italy would facilitate the bombing of Germany and the Rumanian oilfields.

The Americans strongly opposed this. General Marshall wanted all resources to be geared towards a cross-channel operation, any surplus forces not required for the Channel operation; should be employed against Japan.

The American and British conceptions of warfare were completely different. The Americans approached it in the characteristic forthright manner of a nation with unlimited resources, wishing to gel it over as soon as possible. The main enemy was Germany who could be most quickly destroyed by a Channel crossing. Therefore, everything should be devoted to that end.

The British, on the other hand had neither the men nor the industrial capacity to view it in this light. Britain's aim, as a sea-power, has always been to avoid head-on collisions and wear down the enemy. The Americans wanted to out produce Hitler; the British wanted to out manoeuvre him.

After much argument at Casablanca, the Americans admitted, grudgingly, that a cross-channel operation could not be staged in 1943, and fearing that the Russians might come to terms with Hitler, as they had in 1939, agreed to the invasion of Sicily. It was only the subsequent fall of Mussolini that satisfied the Americans that this decision was correct.

A further source of disagreement between the Allies was the 'unconditional surrender' call, which was Roosevelt's idea. Had it been differently phrased for Italy, Mussolini might well have paid some attention to it. But this demand, together with Allied propaganda depicting him as the sole agent of Italy's disasters, drew him closer to Hitler. Churchill was in favour of treating Italy differently from Germany, but was finally persuaded to agree with Roosevelt that a more lenient treatment towards Italy may antagonize their Balkan allies, Greece and Yugoslavia.

The choice facing the Allies was whether to attack Sardinia for a landing near Rome, southern France, Sicily or the Balkans. After a Balkan and French invasion had been ruled out, the Allies chose Sicily, but persuaded the Axis of Sardinia.

On May 9th 1943, the corpse of a British officer was found washed up on the coast of southern Spain, attached to his wrist was a brief-case containing the Allied order of battle for the Invasion of Italy. The Spanish authorities handed this over to the Germans, who learned from it that the Allied invasion was to start in Sardinia, combined with a feint attack on Sicily. To the German general staff this was logical. The notion that the Allies would work their way up a mountainous peninsula, more favourable to defense, was illogical and the deception confirmed this diagnosis.

The German high command issued orders for the defense of Sardinia to take precedence over Sicily. Since they had agreed to seek ultimate victory in Europe by means of a cross Channel operation, was any other Mediterranean undertaking feasible? Because the available resources precluded simultaneous major campaigns in both areas, Mediterranean ventures would draw off the build-up being accumulated in the United Kingdom for the cross-Channel attack, and would probably postpone the decisive action. Yet the southern shoreline of Europe between Spain and Turkey, occupied by Axis troops, was close at hand and a tempting target for invasion. Was it, then, better to halt after Sicily and conserve resources for a quicker thrust into north-west Europe, or to employ against the European underbelly the men and material gathered in the Mediterranean?

The Americans, more conscious of the war in the Pacific, inclined toward an early cross Channel invasion, while the British, who generally regarded a cross-Channel operation as the decisive strike against a weakened Germany, favoured continuing the war In the Mediterranean. As it became clear during the spring of 1943 that shortages of assault shipping and the German strength would prevent a cross Channel effort that year. Mediterranean ventures beyond Sicily became increasingly practicable and attractive.

Where to go? The Americans generally looked toward the western Mediterranean, with conquest of Sardinia and Corsica leading to an invasion of southern France in order to complement a main effort across the Channel. The British generally looked to the eastern Mediterranean, with landings on the foot of Italy leading to an invasion of the Balkans in order to support the Yugoslav Partisans, draw Turkey into the war, and open a shorter sea route to the USSR for lend-lease supplies. Both courses of action had serious disadvantages.

One hope intrigued the Allies: if invasion of Sicily failed to force Italy, the weaker European Axis partner, out of the war, would an additional blow do so? If Italy surrendered, the Germans would be doubly stretched over the periphery of Europe; they might even withdraw from Italy, thereby giving the Allies airfields for intensified bombing attacks. But if the Germans chose to fight in the rugged Italian terrain, they might provoke a protracted campaign that would probably require the Allies to increase the resources in a theatre relegated by the Combined Chiefs to subsidiary importance.

When the Allied leaders met in Washington for the Trident Conference during May 1943, as the North African campaign was ending in triumph, they confirmed their plans to invade Sicily, scheduled the operation for July, and decided on their Mediterranean goals: they would try to knock Italy out of the war, and at the same time engage and contain the maximum number of German forces. But when they tried to define how to attain these aims, they could reach no agreement.

To find a specific solution, Churchill, General Marshall, the US Army Chief-of-Staff, and General Sir Alan Brooke, Chief of the Imperial General Staff, travelled to Algiers to meet with the Mediterranean commanders: General Eisenhower, the Commander-in-Chief, Allied Force; General Alexander, his deputy commander; Air-Chief-Marshal Tedder; and Admiral Cunningham.

The commanders concluded that heavy pressure during the next few months might well force the Italians to surrender. But the choice of operations after Sicily would have to depend on two Axis reactions impossible to know in advance: whether the Italians would disintegrate during the Sicily campaign, and what the Germans would do about continued Italian demoralization. Eisenhower felt that if conquest of Sicily failed to eliminate Italy from the war, the Allies ought to go directly to the mainland; Churchill agreed.

The easiest way to get on to the mainland and into southern Italy was to cross the Strait of Messina. In June, a month before the invasion of Sicily, the British Chiefs-of-Staff began to see a campaign in southern Italy as far north as Naples, even Rome.

For months Allied general staffs planned the largest amphibious assault on an enemy held coastline to date.

When the Allies invaded Sicily, it soon became clear that this was not a feint attack, and that the Germans had misjudged the situation entirely.

The key objective of the invasion of Sicily was the town of Messina to cut off the flow of German and Italian reinforcements from the Italian mainland only three km's across the straits. Messina, however, could not be attacked direct as it was outside the range of Allied fighters.

The experience at Dieppe had shown that a direct attack from the sea against any defended port would be costly. Accordingly it had been decided that future attacks upon ports were to be made by troops landing on beaches out of range of the port defences. After disembarking, the troops would fight their way overland to take the port from the flank or the rear.

Off shore bombardment would be provided by six British battleships- Nelson, Rodney, Warspite, Valiant, Howe, and King George V.

Before the landing began, one obstacle had to be removed from the path of the invasion forces, the Italian garrison on the island of Pantelleria. At this stage of the war the tiny island still had a tremendous reputation, according to Fascist propaganda; it was a second Gibraltar or Malta.

Therefore an entire British iinfantry Division was sent against Pantelleria on June 11th 1943. But as soon as the force arrived, the island garrison promptly surrendered, on the pretext that there was no water available. This turned out to be untrue. The unwillingness of the garrison to make a stand precipitated coming events. With Pantelleria in Allied hands, the seaborne assault on Sicily could proceed on July 10th.

The armada consisted of two thousand five hundred ships and landing craft, carrying one hundred and sixty thousand troops, fourteen thousand vehicles, six hundred tanks, and one thousand eight hundred guns, under an air umbrella of more than four thousand aircraft. The only part of the Allied landing plan to go astray was the airborne thrust to seize certain key bridges and road junctions. Inexperienced flying crews dropped the paratroops and gliders all over southern Sicily. Nevertheless, small groups of paratroops played a vital part in delaying German counterattacks which could have destroyed the US beach-head. Despite lack of training in amphibious landings, the Allied troops achieved complete surprise and went ashore smoothly, aided by the new amphibious truck, the DUKW.

Once the initial German counterattacks had been broken, Alexander, overall commander of the ground forces, envisaged the British 8th Army thrusting up the eastern coast of the island to seize Messina, and to cut off the Axis forces before they could, retreat to the mainland.

But the plan misfired; stubborn German resistance prevented the 8th Army from breaking out. Meanwhile Patton's forces swept across the island to Palermo in a classic armoured thrust, and then worked on along the northern coast to reach Messina before the British.

Until El Alamein, most Italians thought that their country might emerge from the war victorious. But after a series of defeats, coupled with Stalingrad, discontent with Mussolini now appeared openly. As military affairs went from bad to worse, a number of anti-Mussolini groups began to emerge in Italy, among them was the King, who believed the war was lost.

He had become aware too, that Germany was beginning to regard Italy less as an ally than as a bastion, to keep the war as far away from Germany as possible. He believed at this juncture that a direct approach should be made to the Western Allies, attaching much importance to the belief that the Anglo-Americans did not wish Russian influence to extend to the Mediterranean. Here, he appears to have underestimated the solidarity of the anti-Axis alliance. The west had no intention of jeopardising good relations with Russia.

Once Sicily had fallen, the Allies planned;

1. To cross the Straits of Messina,
2. To land near Naples to draw off German forces.
3. To seize the valuable port of Taranto.

All forces would then join in a concerted drive to Rome. But the Allied planners were unable to anticipate the ability of Kesselring, German C-in-C Italy, to produce a formidable defence.

When the Allies decided to invade Italy rather than southern France or the Balkans, they hoped for an easy task. They thought that the Italians would not fight and the Germans could not transfer sufficient troops into Italy to impede an Allied drive to the north. But they were to receive a severe shock. Italy proved to be no 'soft underbelly." Despite success at Messina, the landing at Salerno was a most desperately and close run battle.

When the relatively easy invasion of Sicily showed the extreme deterioration of Italian military power, the question arose whether a landing on the west coast of Italy near Naples was more feasible than Sardinia. On July 18th, Eisenhower approved the carrying of the war to the Italian mainland, and on July 25th, King Victor Emmanuel removed Mussolini from power.

The Italian government under Marshal Badoglio, while assuring the Germans that Italy would remain in the war, was negotiating secretly with the Allies to surrender. During the surrender negotiations the Italians offered to open to the Allies the ports of Taranto in the heel of Italy, and Brindisi on the east coast. The few Germans in the region were expected to withdraw; the Allies would seize these excellent harbours to facilitate a build-up in Italy. They would have two lines of communication, one based on Salerno and Naples for the US 5th Army, another based on the other side of the Italian peninsula for the British 8th. Eisenhower decided to move troops on into Taranto as soon as the Italians capitulated. The codename 'Slapstick' Alexander later remarked, well-illustrated the makeshift nature of the planning.

Would the Germans fight to repel the landings, or would they retire to the north? The Germans were suspicious of Italy's intentions. If Italy collapsed, Hitler would have to fill a vacuum in the Balkans and in southern France where Italian troops occupied coastal regions. He would have to assume the defence of all of Italy, or withdraw to the Alps, or hold the rich agricultural and industrial resources of the north. His earliest thought was to occupy and defend the entire country, and in May 1943, appointed Rommel to command Northern Italy. But operations in Russia made it impossible to supply Rommel with adequate forces.

While Hitler and Rommel made plans in anticipation of an Italian defection, Kesselring, the senior German officer in Italy, worked in close cooperation with the Italians to turn back any Allied invasion. Shocked by Mussolini's fall from power, he accepted in good faith Badoglio's declarations that Italy would stay in the war. Not so Hitler, who decided in August to occupy Italy by sending additional, forces into the country, ostensibly to augment the defences. If the Italians capitulated to the Allies, Rommel was to move into northern Italy while Kesselring pulled his forces out of southern Italy.

In August, increasing numbers of German divisions entered the mainland of Italy as Rommel moved troops in from the north and Kesselring withdrew the units that had fought in Sicily. The Italians felt insecure because they had been unable to reach a surrender agreement with the Allies. They had no doubt that the German troops in the north were an occupation force, but, too weak to protest, they pretended to accept the German explanation that Rommel's units were a strategic reserve.

By the end of August, Hitler had reached a firm decision. If the Allies landed in Italy or if the Italians turned on the Germans, Kesselring was to withdraw to the Rome area, holding there only until the troops in the south had safely retired and those on Sardinia and Corsica had reached the mainland. Then he was to move to the north and join Rommel's forces. At the beginning of September, Allied strategic bombing attacks had forced the Axis to remove their aircraft from all the major fields in southern Italy.

On September 3rd, four years to the day after Britain had gone to war, Montgomery's 8th Army, with massive artillery, naval, and air support crossed the Strait of Messina and invaded Italy. There was no opposition; some Italian troops even volunteered to unload the landing craft and ships. To get ashore and take Naples was the mission of Lieutenant-General Mark W. Clark, who headed the US 5th Army and who would have his first battlefield command in the Second World War.

Once the news of the Italian capitulation was broadcast to the troops on their way to the beaches of Salerno, few men in 5th Army anticipated much trouble. But they had bargained without Kesselring's lightning deployment of his troops. The Allied troops established themselves on the beaches, but the first attempts to push inland were thrown back by devastating German fire. The Allies expected the Italians to join them in fighting their former allies, but the Italian rank and file considered themselves sold out and betrayed when told that the Royalty and government had fled post haste from Rome to seek shelter in the Allied controlled south.

The Germans in the Salerno sector launched such a deadly counterattack that they nearly drove clean through 8th Army to the sea. As it was, Clark's US 6th Corps and British 10th Corps were separated and tied down by German attacks all along the Allied front, while Clark himself almost decided to cut his losses and re-embark 5th Army. But the front held, reinforcements poured into the beach-head, and Kesselring could not ignore the steady advance of 8th Army in the east. German strategy demanded a withdrawal, regardless of the outcome at Salerno, and after a ten-day crisis for the Allies, the Germans fell back. As they did so, Clark's forces broke out of the beach-head, driving north-west towards Naples

Hitler and Kesselring were satisfied, they had denied the Allies quick access to Naples, inflicted severe losses on Allied units, extricated their forces from southern Italy, and prohibited the Allies from fully exploiting the Italian surrender. The Germans may claim with some justification, to have won an important success with limited forces. The Germans had sustained casualties of three thousand five hundred men. In contrast, the Americans lost about three thousand five hundred, the British five thousand five hundred.

The invasion of southern Italy caused the Germans to abandon Sardinia and Corsica. By virtue of their strategic location, the islands were a great prize and made the Mediterranean even more secure, particularly Corsica that would place Allied bombers closer to the German homeland. When Kesselring gave permission to break off the battle at Salerno, he ordered a slow retirement to the north to gain enough time to fortify *a* naturally strong line about fifty miles north of Naples and ninety miles south of Rome. The terrain around Cassino provided even better prospects for a prolonged defensive battle.

While withdrawing, the Germans destroyed all installations of military value in order to deny their use to the Allies. By the end of September he had extended a front across the peninsula of Italy. The US 6th Corps under a new commander, Major' General Lucas, moved from the Salerno beaches into the interior, that was more than adequately defended by small rearguard units employing demolitions and mines with skill to impede progress.

The cost of the twenty one day campaign was more than twelve thousand British and American casualties. The Germans pulled back to the prepared Gustav Line. And it was around this time that Churchill abandoned any hope of marching into Rome before the end of 1943.

Immediately after the capture of Naples the Joint Chief s-of-Staff in Washington urged that Major-General Doolittle's strategic bombing force should be moved from North Africa to southern Italy, setting up air bases for attacking targets in south Germany and Austria. Bomber attacks on the ball-bearing works in Turin, the railway junctions of Genoa, Bologna, and Innsbruck, and the Rumanian oilfields of Ploesti and other crucial strategic targets.

From the spring of 1944 onwards massive bomber raids on the aircraft industries in southern Germany and the oil supply system from the Balkans were particularly hard hit. In the summer of 1943 Germany had still been drawing two hundred and twenty thousand tons of oil a month from Rumania and Hungary; by June 1944, the oil imports were down to forty thousand tons. The setback to aircraft production and the impact of the oil shortage on all the services greatly affected Germany's defensive preparedness for the impending D Day invasion, as well as reducing her striking power in Italy and on the Russian front.

In November 1943 the British 8th Army started its offensive against the left flank south of the Gustav Line with a powerful artillery barrage by nine hundred and twenty five guns. It was the strongest concentration of firepower yet assembled in the Italian campaign. No less than two hundred and seven thousand shells were discharged. The battle raged to and fro and by December 6th the Allies breached the southern pillar of the German line.

The attack on the northern pillar was launched by the American 2nd Corps on December 8th. In this assault an Italian brigade also took part, the first Italians to be sent into battle against German troops since Italy's December 28th declaration of war on Germany. But this brigade, together with the US 36th Division, was foiled by the resistance of the Panzer Grenadiers and it was not until December 16th that the Americans and Italians, after an extremely heavy artillery barrage, took the position. Bit by bit, the German XIV Panzer Corps was forced back to the Gustav Line.

By December 1943, Alexander's drive to Rome had been completely halted before the defences of the Gustav Line. Originally, Hitler favoured Rommel's plan but on November 21st he transferred Rommel to France, leaving Kesselring as C-in-C Army Group Italy. The clash between Rommel and Kesselring, which dated from the desert war, thus resulted in a triumph for Kesselring; a triumph soon to be confirmed in battle. With Corsica and Sardinia, evacuated and the allies established in the south, Kesselring was determined to deny Rome and its airfields to the Allies. By the end of 1943 Kesselring's Gustav Line defeated every breakthrough attempt.

In theory, the Allied invasion of Italy was intended to act as a gigantic magnet, drawing off as many Axis divisions as possible from the Channel coast and the Eastern Front. But Rome exerted a hypnotic fascination over the Allied Command. The Allies faced the problem of whether to continue up the Italian peninsula, and if so, how far. The Combined Chiefs-of-Staff directive stated simply that General Eisenhower, the Supreme Allied Commander, was to eliminate Italy from the war, already accomplished, and to contain the maximum number of German divisions. Without specified objectives, the Italian campaign, according to General Alexander, the Allied ground forces commander, became, in retrospect, 'a great holding attack'.

Yet objectives gave purpose to operations and determined the resources required. Before the invasion of Sicily, planners assumed that the Germans would not reinforce a collapsing Italy and would, after an Allied landing on the Italian mainland withdraw to the northern Apennines. Nothing but Rome, wrote Churchill, could satisfy the requirements of this year's campaign.' The quick conquest of Sicily led the Allies to expect a rapid advance to northern Italy and, from there, overland and amphibious operations against southern France and Balkans

The taking of Naples after the Salerno landings, pointed toward Rome. A static situation failed to tie down thc maximum number of German troops. Either the Allies would drive the Germans out of Italy or be driven out themselves. The Combined Chiefs who were meeting in Quebec for the Quadrant Conference, anticipated that Allied troops would have to fight their way slowly and painfully up the peninsula.

The Combined Chiefs stressed 'the importance of securing the Rome aerodromes'. General Marshall, who had previous reservations on the overall value of an Italian campaign, agreed that Rome ought to be seized as quickly as possible. On October 1st Eisenhower expressed hope of being north of Rome in six or eight weeks. Three days later, he and Alexander expected Allied troops to be in Rome within the month. Eisenhower planned moving his headquarters from Algiers to Naples but decided to wait until he could go straight to Rome. Then all expectations that the Germans would fight only delaying actions in southern and central Italy underwent a startling change when German divisions were observed moving from northern Italy to reinforce the troops south of Rome. There would now be very hard and bitter fighting before Rome is reached.

Alexander judged that the Germans had recovered from the Italian surrender and now held a numerical advantage in ground troops. The

Germans had been retreating since November 1942, from El Alamein, across Libya, Tunisia, then out of Sicily, and finally out of southern Italy. It was time to stop for the sake of troop morale alone. But there was now a political reason also. The Germans had rescued Mussolini and had established a government under his authority, so retaining Rome as Mussolini's capital would strengthen the semblance of his restored power.

Whatever the reasons, the Allied theatre commanders were convinced by mid-October that operations would meet progressively stronger resistance. Yet requests for additional resources received a negative response. Seven divisions were to be withdrawn for participation in the cross-Channel attack. These divisions were to be replaced by French formations being re-equipped and trained in North Africa.

The German command considered that a prolonged defence in southern Italy would delay an Allied invasion of the Balkans, keep Allied bombers farther from Germany, retain Rome; and require fewer troops than the line in the northern Italy. Hitler instructed Kesselring to hold at the Bernhard Line on the narrowest part of Italy whilst construction went on to fortify the stronger Gustav line and Monte Cassino.

To resolve the conflicting strategies of his commanders, Hitler called them to a conference on September 30th and listened to their views. Rommel appeared pessimistic, while Kesselring, optimistic, promising to hold the Allies in southern Italy for six to nine months'.

While the US 5th Army under General Clark clashed with Kesselring's strong and determined forces on the western side of the Italian peninsula, Montgomery's regrouping was so time-consuming, that the Germans were allowed not only to retreat to a strong defensive position but also to strengthen their fortifications further back on the Sangro river. The US 5th Army's troops were finding that despite their overwhelming superiority in aircraft and artillery, the less well-equipped Germans were difficult to dislodge. The crossing of the Volturno was a victory for the Allies but the delay in fighting there was also a strategic success for the Germans who desperately needed time for work on their main Italian defence line.

In December 1943 Montgomery was transferred to Britain to prepare for D Day and handed over command of 8th army to General Leese. In January 1944 the Allies resumed their drive on Rome, and their new offensive met with the same fate as that of October/November. German resistance blocked the Allied attacks, and the Allied forces had come up against the strong German fortifications of the Gustav Line, centered on Cassino

The Germans had been pushed back in the Adriatic sector near Mignano. But the mountain warfare had been so grueling that the Allied units were depleted and exhausted. The difficulty of advancing across the rugged terrain in southern Italy had long prompted the thought of an amphibious movement. Sending troops by sea to land behind the German front might loosen the defences and bring about more rapid and less painful progress.

Yet serious obstacles interposed. In addition to the direct deterrents erected by the Germans, offshore minefields and strong coastal defences at suitable landing sites, the Allied forces lacked sufficient troops and ships for a substantial amphibious venture. The Allies could make only small and shallow amphibious operations in Italy, and these would hardly be decisive in changing a virtually static campaign into mobile warfare. Rome appeared to be out of reach in the immediate future.

By January 1944, the Allied generals at Cassino were ready to launch their first determined attack to break through the Gustav line, join up with a projected landing force to the north at Anzio, and take Rome. But the clumsy frontal attacks were all repulsed with murderous losses and by mid-February the Allies were forced to admit defeat and break off the battle.

~TWENTY THREE~
BATTLE OF THE BULGE

With the Allied armies poised to invade Germany towards the end of 1944, Hitler became convinced that his only hope of turning the tables and saving the Third Reich was to concentrate the last of his reserves in the west. He needed no reminding that in five years of war the German armed forces had lost almost four million men, the elite of German manpower. Plus, the essential raw materials from Russia, Finland and France were no longer available, and, devastating thousand bomber raids on German cities had become common place. Yet Hitler still saw hope, or professed he did.

In this momentous struggle for survival, the Germans had no option other than offensive action in a desperate attempt to halt the tide of defeat. By hitting the enemy hard with one swift pulverizing blow, the entire strategic situation would be transformed. Hitler, therefore, conceived a bold and imaginative military plan which outlined a blitzkrieg across Belgium, with the ultimate objective of re-capturing the strategic port of Antwerp; thereby seizing the initiative once more.

He calculated that his enemies in the west could be defeated and induced to accept a negotiated separate peace, despite their decree of unconditional surrender. Hitler firmly believed that by staking everything, on this massive blow, it would cause the inevitable collapse of the artificial wartime coalition between the ultra-

Capitalists in the west and the ultra Marxists in the east.

Taking into consideration, at this stage of the war, the Wehrmacht's limited capabilities, Hitler's plan was, to say the least, extremely ambitious. But, if successful the gains would be enormous. It would effectively cut off more than half the sixty four Allied divisions engaged in Europe at that time.

At first glance, Hitler's audacious proposed counter stroke appears to be ludicrous. But, as in all his schemes it contained elements of strategic inspiration. Antwerp's port facilities were vital to the Allies whose advance was grinding to a halt. They had outrun their supply lines from French ports which were far from the front. Furthermore, the loss of Antwerp would set the Allied operational plans back by at least a year. Thus allowing the Germans time to complete the development of their new wonder weapons, namely:

V1 and V2 rockets
U - Boat snorkel
Jet aircraft, and
Most importantly perhaps even the Atomic Bomb.

The German generals had serious misgivings about Hitler's plan, considering it to be beyond their military capacity, therefore, impracticable. They did, however, acknowledge that in order to prevent the Third Reich from being totally destroyed in less than six months, it was essential to go over to the offensive, and a last ditch effort on the western front was the only place where this might be possible.

The generals were fully aware that they could not muster such relative power as Germany had had in 1940, particularly in the air. However, it was put to them that the winter weather conditions would compensate for this. There was indeed method in Hitler's madness, for to stand beleaguered on the defensive while his enemies gradually strangled Germany offered no solution. He was, therefore, compelled to go over to the offensive and attempts to strike decisive blows which may alter the entire situation. Hitler, like his generals, realised that there was no hope of a conclusive result being forced against the Soviets in the east. That is while he was still engaged in the west.

He envisaged a "new Dunkirk" situation developing, with the British once again evacuating to England or this time being completely destroyed. The German dictator, therefore, was convinced that a successful offensive in the west would, at a stroke, alter the direction of the war. The defeat of Germany's enemies in the west would stabilise the front and restore the balance of power in Germany's favour.

By inflicting a catastrophic defeat on the western allies he was sure that the Americans, on finding themselves isolated and without support would feel obliged to turn away from Europe to concentrate on the enemy in the Pacific. After all it was Japan who had tarnished American honour at Pearl Harbour, not Germany. On achieving a separate peace, all German strength could then be focused in the east to confront the invading Soviets. Hitler insisted that his forces must go all out to inflict on Eisenhower a defeat as crushing as that suffered by the French in 1940. Therefore, alternative proposals, presented by his staff officers (whom he mistrusted), for more limited objectives were entirely rejected. This would only offer Germany a short respite when a decisive military victory was required to save the Reich.

Hitler ordered the best units to be drained from the eastern front and the remaining German reserves to be deployed to strike against the American army whom, Hitler contemptuously described as merely bank clerks and hoodlums, or, the Italians of the western alliance.

It was accepted that mounting such an offensive must entail enormous risks. Nevertheless, it was absolutely vital that conditions had to be created in which the Allies would be at loggerheads, and blame each other for the forthcoming disaster; thus sowing mutual distrust which would perhaps lead to a split and eventual collapse of the alliance.

Adolf Hitler therefore set in motion preparations for a battle that was to assume epic proportions; the greatest German attack in the west since the campaign in 1940 when he had brought down his enemies in ignominious defeat. This offensive was later to be known as; "The Battle of the Bulge". It was destined to involve more than one million troops, and it would precipitate an unparalleled crisis for the Allied armies by demonstrating the weakness of American battlefield intelligence at that time.

The Supreme Allied Commander, General Eisenhower never imagined for one minute that Hitler would fix Antwerp as the objective for his panzers. On this issue he was of course in total agreement with the German generals. The possibility of a German attack against Allied positions was at no time ever contemplated. The war, according to the Allied commanders was already won and virtually over, victory was assured, it was now only a question of tidying up. As late as December 12th, only four days before the offensive began, Eisenhower was confidently informed by his intelligence staff that the Germans were in no position to stage a major offensive operation.

The over confident Allied commanders, being flushed with success, negligently assumed that the German troop concentrations in the Ardennes region were defensive preparations. Even though reports were received of German activity in the Ardennes, no suspicions were aroused or alarms raised, the reports were largely ignored or misinterpreted. Despite the compelling evidence of the First World War, and of course four years earlier, when the world was staggered by a surprise German armoured blitzkrieg through the Ardennes. The Allied staff officers inexplicably still believed that this region was impassible to tanks.

This complacency and disbelief allowed the Germans by mid-December to assemble twenty five divisions, ten of them armoured. The Allies, blinded by overconfidence, had become victims of their own propaganda, and failed to penetrate Hitler's masterful deception scheme. This was an intelligence failure of the highest order.

Completely unnoticed over a front one hundred and twenty five kilometers long and under an enemy controlled sky, the Germans deployed a formidable force of two hundred and seventy five thousand men, two thousand heavy artillery pieces and one thousand armoured vehicles. They were all in position ready to strike the opposing American lines consisting of eighty three thousand men, four hundred artillery pieces and four hundred armoured vehicles. The Allied supreme command had taken a great risk in manning this porous and thinned out sector by defending it with untried and battle weary troops requiring rest and refitting. Thus, because they had failed to reckon with Hitler's megalomania, the Allied chiefs were to be caught badly napping. The exposed, flimsy and strung out American defence lines were totally unprepared.

The logistical achievement of the German general staff, in concealing the attack displayed a remarkable skill. The German military still possessed a great talent in tactical planning. The Ardennes did indeed seem to the Germans to provide an opportunity comparable to the one exploited in 1940.

At 05 30 hrs on the morning of December 16th 1944, the peaceful illusion along the front was shattered when two thousand secretly massed German guns promptly opened up with a terrifying barrage to stun the troops on the receiving end. Rail guns were also used to good effect, in this the Wehrmacht's last great counterstroke.

When the hour long bombardment lifted, the first waves of German troops then rolled forward. They advanced almost unhindered through the early morning fog and mist, seeking lines of least resistance for the panzers to exploit. They were playing for high stakes on very limited funds.

The offensive was deliberately timed to coincide with weather conditions of low cloud which kept Allied air forces grounded. Thereby, denying the defenders close air support and reconnaissance information of German troop movements. In some sectors the shattering psychological blow simply overwhelmed the front line American defences to create panic, confusion, and disorganised retreat. Few areas managed to coordinate an orderly withdrawal. Surprise had been achieved on a grand scale and the poorly supported and overextended American lines were taken completely off guard and off balance. The fearsome German Tiger tanks advancing with machine guns blazing or 88 mm. guns blasting presented a paralysing sight.

The Germans continued to probe seeking softer targets, only occasionally meeting pockets of Americans in isolated units who held firm putting up strong resistance from frozen foxholes. But despite desperate defending the situation deteriorated rapidly for the Americans who could not prevent the enemy penetration towards the crucial road juncture at Bastogne.

After two days of severe fighting the pattern of the battle was becoming clearer to the Allied commanders. Realising the implications and the magnitude of the threat, there was little room for complacency. The German advance was no longer regarded as a spoiling attack aimed to force the abandonment of a likely allied offensive. German pressure was intense almost everywhere, despite stiffening American resistance, combined with precarious supply lines and enormous traffic jams. The Germans made gradual progress and achieved the elusive breakthrough.

The panzers were assisted by the confusion sown in the American lines by a brigade of English speaking German commandos in American uniforms, whose mission was to infiltrate American lines and create maximum confusion. The German spearhead tank assault was directed towards the river Meuse and was headed by the SS panzer division "Liebstandarte Adolph Hitler". This elite force was commanded by veterans of the eastern front experienced in taking full advantage of the element of surprise. Nothing now seemed to stand between the Germans and the river Meuse, where they intended to secure the bridges in order to launch the main attack towards Antwerp, their ultimate objective.

They were, however thwarted and not by massively superior forces but by small groups of American combat engineers. The Americans effectively blocked roads, blew bridges and destroyed fuel dumps to slow down the panzer advance. The initial German gains of surprise and speed were slowly fading stopped both north and south. The Germans were being channeled into a salient resembling the jaws of a trap.

It may be of interest to note that German paratroops could have been used extensively and effectively to seize strategic bridges. But Hitler had allowed this specialist arm to dwindle and decline after suffering major losses in 1941 at Crete against the British. Therefore only a few small groups were used piecemeal and played an insignificant role.

The German attack towards Bastogne initially fared better, their objective had been reached as planned and the American lines were breached. However, Bastogne was destined to become a thorn in the side of the German commanders who were by this stage fully aware of its strategic importance. They realised it held the key to victory as General Mcauliffe's American 101st airborne division garrisoned at Bastogne could launch counter attacks against German forward positions. Arriving at the same conclusion, Eisenhower ordered Patton, the controversial and flamboyant American general to suspend current operations and wheel 90 degrees; shifting his troops two hundred and fifty kilometers northwards to strike through the German flank and reinforce Mcauliffe at Bastogne.

Patton manoeuvered with impressive speed and skill; he had a tremendous capacity for organisation and was a planner of the highest class. But, before Patton's Third Army could intervene the panzers broke through the Bastogne outer defence ring. The Americans were offered surrender terms on December 22nd. This caused high drama with Mcauliffe's inspired and famous reply **"NUTS".**

This is probably unequaled in the annals of military history. Both for its brevity and the amount of confusion it caused in the enemy camp before it was deciphered. It was a reflection of the growing confidence among the defenders and indicated the garrison's high morale. The German time tables were being totally disrupted, at Bastogne, giving the Allied commanders a few vital days to consolidate.

That same day the skies cleared and the Allied air forces were immediately unleashed to airdrop supplies into the shrinking perimeter of the beleaguered Bastogne garrison. Allied air power now played a decisive role in the battle creating widespread havoc behind enemy lines. Two thousand American bombers escorted by five hundred fighters actually attacked German airfields and communication networks as far distant as Frankfurt.

On Christmas day the Germans mounted a ferocious air and ground attack against Bastogne. The fighting was bitter but the American paratroopers held on grimly until Patton's relief force broke through to link up with the defenders. The German attack ground to a halt. Not only was Bastogne saved, but Patton's armour had torn a huge gap through the panzer formations threatening them with encirclement and isolation. For the Germans, it had by now become a matter, not of Antwerp but of survival. The German leader, however, refused to accept the reality of the situation and the hopeless position of his troops. The advance was not to be abandoned. He had been informed about the fierce rivalry within the Allied camp which he believed could be exploited.

He insisted that Allied efforts were uncoordinated and urged his generals to eliminate Patton's corridor to Bastogne. This would render an advance on the Meuse and beyond still possible and would further exacerbate problems within the Allied command. The allegations of disunity between the Allied commanders were not entirely unfounded. A crisis had arisen which highlighted the seriousness of the disagreement between British and American opinion on how the war should be finished. Montgomery revived his demand for a single thrust into Germany, openly opposing Eisenhower's more cautious broad front strategy. Furthermore, he demanded that he should be placed in command of this effort.

Eisenhower, had by this time become sick of the controversy, and, considering there were forty four American divisions in Europe as compared to twenty British, he quite naturally took exception at Montgomery's demand and threatened to sack him. It took enormous diplomatic activity behind the scenes, particularly by Churchill, to soothe over the affair and ease the strains; only to see it blow up again when Montgomery rather tactlessly stated to the press that he had saved the day for the Americans at the Ardennes.

All this wrangling coincided with the continued heavy fighting in the Ardennes, therefore, Eisenhower made clear that he would tolerate no further arguments within the Allied command, particularly between Montgomery, Bradley, and Patton. Montgomery, realising how insensitive he had been, apologised to Eisenhower.

In the meantime, Patton, after receiving further reinforcements managed to broaden the corridor to tilt the balance in favour of the Allies. The Germans did however attempt to rescue the situation by launching a massive air attack on Allied airfields in Belgium and Holland. Over three hundred Allied aircraft were destroyed but the Luftwaffe lost two hundred and fifteen aircraft shot down by the Allies plus eighty five shot down in error by themselves. These losses were irreplaceable. It was a death blow for the Luftwaffe. Hitler's air arm was permanently crippled.

The Allies launched a pincer attack with nearly one hundred thousand men to push back the remaining German forces who continued to fight with great tenacity in retreat. The Ardennes battle was over. The great wedge driven into the front was finally pinched out. It had cost the Allies in which the Americans bore the brunt over seventy five thousand casualties and the Germans one hundred and twenty thousand.

The unexpected and menacing German counter offensive had disrupted the Anglo American plans and inflicted much damage. But, it fell disastrously far short of its objectives and proved to be a fatal operation for Germany regarding the ultimate outcome of the war. Hitler's grandiose scheme may have been brilliant in concept, and it came ominously close at times. However, in retrospect, unlike 1940, the Germans did not have the resources to exploit their tactical success.

As far as American combat soldiers were concerned, they had stood toe to toe with the Wehrmacht on a battleground chosen by Hitler and had finally broken the back of the German army in the west. Hitler's daring plan in trying for the grand slam had failed. The final outcome of the war in Europe was no longer in doubt. Any chance of saving the Third Reich had vanished; the end was now just a few months away.

No conceivable strategy could turn German defeat into victory. The German armies which had seemed irresistible for the first few years of the war were now outnumbered, outgunned and outmatched on every front. Hitler had gambled and lost, and in losing, had not only sealed the fate of his armies in the west, but had also made defeat in the east inevitable.

In the aftermath of the battle, the Allies advanced into German territory in early 1945 and were faced by under equipped and under trained German troops mostly recruited from the Hitler youth or home guard. Hitler had issued a proclamation that all able bodied men aged between sixteen and sixty were to be called up. In truth many of the recruits were over sixty and as young as eleven years of age, a pitiful ragtag of middle aged men and adolescents. German civilians had long ceased to believe in the Nazi propaganda of promised “miracle weapons" that would dramatically reverse the tide of defeat.

After the Allies had swiftly secured the bridgeheads over the Rhine, to trap and destroy the remaining German forces in the Ruhr, both Churchill and Montgomery urged Eisenhower for a bold thrust towards Berlin. The British prime minister was horrified at the prospect of leaving the German capital to be taken by the Russians, regardless of the agreement at Yalta. He considered the Americans to be politically naive regarding the European post war carve-up. However, Churchill's appeal to the dying President Roosevelt for a push to Berlin was in vain. The American president had given Eisenhower full powers of decision. Ironically, this was unlike Truman and MacArthur during the Korean conflict a few years later.

But, Eisenhower was adamant, the main prize was not Berlin, it was the annihilation of enemy forces. He was of the opinion, (unlike his Russian counterpart Zhukov), that Berlin no longer had any strategic importance. Furthermore, the Allied Supreme Commander had also become increasingly concerned about disturbing reports of a German last stand in a "national redoubt" in the Alps. He therefore swung his main force south towards Munich to cut off this imaginary threat, thus, leaving Berlin to the Soviets.

The determined and vengeful Red Army, after weeks of preparations launched a massive assault on Berlin. The German defenders, mainly raw recruits, as previously described put up fierce resistance to inflict severe casualties on the attackers. However, the overwhelming superiority of the Russians enabled them to break through north and south of the capital to complete the encirclement of Berlin on April

24th.

The Soviet troops, many of whom had seen their own towns and villages obliterated by the Germans, had fought their way from Stalingrad to seek this moment of revenge. As final German resistance crumbled, wholesale rape and looting occurred. By the end of April 1945 the German army ceased to exist as a coherent force and the Soviet red flag fluttered over the ruins of the Reichstag.

There has been much controversy over the years, regarding Eisenhower's decision on the Berlin issue. What if the Allies had headed straight for the German capital? Would German resistance have been just as ferocious as it was against the Soviets, who suffered enormous casualties? Some estimates go as high as three hundred thousand troops. This is more than the total allied losses since D Day.

Bearing in mind, the fact that Stalin was equally as mistrustful of his western partners; the Soviet leader may have viewed an Allied attack on Berlin as a breach of the Yalta agreement. Perhaps this would have provoked a conflict between the wartime allies, which of course was what Hitler so fervently desired. The acrimony would have marred victory over the Third Reich.

Hitler committed suicide in his bunker on April 30th. For him there was no other way out. His body was soaked in petrol and set alight. His appointed successor Admiral Dönitz, vainly tried to negotiate a separate surrender with the west. He desperately wanted to continue the fight against the Soviets with a view to rescuing as many German civilians and refugees as possible, who were fleeing in terror from the Soviet invaders.

Montgomery and Eisenhower were unsympathetic, refusing to accept any compromise and demanded unconditional surrender. With their cities ruined and reduced to rubble, the stench of death everywhere and millions of people helpless and hungry, the Third Reich capitulated. The Germans finally accepted total defeat and signed the unconditional surrender documents. This merciless conflagration which had lasted almost six years ultimately came to an end on May 8th. 1945.

The consequences of Nazi rule in Germany and their insanely ambitious policies, was that over fifty million people perished. It is difficult to believe that for twelve years, Hitler and his group of fanatical henchmen had guided Europe's strongest nation. In 1940 the world which was now closing in around their heads, had laid at their feet. The great paradox is that Adolf Hitler had raised Germany from the humiliation of Versailles and formed an army which proved capable of carving an empire surpassing even that of Napoleon. The German army had won many great battles and its generals were acknowledged as the finest thinking and fighting soldiers in modern history. But, the political decisions of Germany's leaders made victory impossible.

Now came the epilogue, Hitler was dead, but, the generals who had followed him under the swastika as willing executioners of his policies had survived, only to face retribution. Inevitably they rounded on their former hero; no longer seeing him as an inspired genius, but as a contemptuous, incompetent amateur whose blunders lost the war for Germany and propelled their country across all bounds of civilisation.

There have been many crimes committed against humanity in recent times, but the pitiless cause the German military with all its savagery and barbarism has no parallel. The mind boggles at their claim that they did not know anything of the unspeakable crimes, and in any case, they were merely following orders; thereby, dispatching the guilt entirely on their political masters. The question arises, were they more monstrous than the scientists who created the atom bomb? Is the comparison relevant?

Was the allegiance the western scientists displayed to their cause not the same as Hitler's generals gleeful embrace of theirs? This is a debate that could continue into eternity! Whatever the conclusion, there is one undeniable and absolute fact! Hitler's generals would never know what it would be like, to live in a world, had they been victorious!

~TWENTY FOUR~
THE ECLIPSE OF THE LUFTWAFFE

In the final stages of NAZI Germany's air defence, some six months before the end of the European phase of World War Two, Luftwaffe fighter pilots sacrificed their lives in a last desperate effort.

In an unequal challenge, they took off time and again to contest Allied fighter escorts that completely outnumbered them, and with the German defence perimeter shrinking, the air war entered a ferocious phase. The heavy losses suffered by Allied air forces, despite their superiority during the last six months of the war is testimony to the bravery and ingenuity displayed by Luftwaffe airmen in their frantic endeavours. In total more than eleven thousand Allied heavy bombers were lost in the European theatre of operations. In fact, after almost six years of continual fighting on several fronts, the once powerful and feared Luftwaffe was reduced to only a token force, and the mighty Luftwaffe which in earlier years had dominated the skies above continental Europe was no longer able to halt the Allied formations pounding their Fatherland.

After September 1944, the Luftwaffe did not have the resources to counter the Allied strategic bombing campaign. They were by this time incapable of inflicting the serious losses on Allied aircraft, armour and troops they had previously achieved. The tide had turned, especially with the development of Allied long-range fighters to escort the bombers deep into Germany.

One could take the view that the eclipse of the German fighter force was due to Hitler's quest for an empire in the East; especially, considering the enormous losses suffered at Stalingrad and subsequent Russian winters.

The entry of the United States into the war, plus the resurgence of the RAF's offensive power slowly eroded the Luftwaffe's strength in a war of attrition. By mid 1944 it had virtually disappeared from the skies over occupied territory; thus leaving the German Armies to fight without air support.

Göring, refusing to accept the true situation, boastfully asserted that "his Air Force" would soon fix things. But when the setbacks came; he accused the pilots of incompetence and cowardice, and had lost mastery of the air through their own fault.

However, from its inception the Luftwaffe was developed for an offensive strategy, to support the ground troops. In the light of early campaigns this decision may have looked sound enough. No one at Luftwaffe high command gave any thought to the course the war might subsequently take. The lessons of the Battle of Britain were not heeded, and offensive thinking remained in the ascendancy. Meanwhile, the Allies, having formed a correct appreciation of the likely turn of events, were able to build up powerful strategic air forces.

By the time the Luftwaffe High Command realised the danger, the fighter force, was suddenly thrown onto the defensive. They had no choice but to spread out in penny-packets and were short of manpower, fuel and equipment. Insufficient attention was paid to aircraft production until Speer took charge. Under his management, production of fighter aircraft reached its peak in September 1944. But, this proved to be too little too late due to German losses, inadequate training of pilots and overstretched resources.

The Me 262 was the world's first operational jet fighter that had its first flight tests in July 1939. The Luftwaffe immediately saw its advantages and called for immediate production. The aircraft was demonstrated to Hitler on November 1943. To everyone's surprise he immediately saw the Me 262 as a fast bomber.

Despite protestations, Hitler remained adamant. He demanded that the Me 262 be developed as a bomber; a role completely alien to its concept. His obstinacy delayed the aircraft coming into service and led to its misuse when it finally did reach operational status.

The advanced nature of the Me 262 resulted in further delays due to inadequate training. In many cases pilots made only two training flights before going into action.

In April 1944 the first Me 262s underwent their first operational sorties and quickly claimed successes. As 1944 drew to a close, many Luftwaffe bomber units were disbanded and re-equipped with Me 262 fighters.

The Allies soon realised that the most effective, indeed, almost the only counter measure against this aircraft was to destroy it on the ground. They placed standing patrols above their airfields to shoot them down when they took off or came in to land

The jet fighters were mainly used to attack four-engined Allied bombers - a tactic that many German fliers considered wasteful. They believed the jet should engage the American fighter escort, leaving the German piston-engined planes to attack the bombers without hindrance

By now, the Allies recognized the Me 262 as a real threat, and considered that one Me 262 was the equal of eight Allied fighters. But however good its performance, it could not succeed against overwhelming odds.

Allied escort fighters were by now thrusting deeper into German occupied territory. The Allies also started to launch raids on Austria and South Germany from Italy; this opened up a new front on which defences had hurriedly to be set up.

In these circumstances the omens for the defence of Germany were gloomy enough. At the beginning of 1944 the new escort fighter, the

P-51 Mustang, arrived in England, and as early as February, wide-ranging air battles developed over North Germany.

In June things were comparatively quiet over Germany, all Allied resources being focused on the landings in Normandy. This breathing-space did not however last long, for once Allied troops had gained a firm footing and begun to advance, the air war over Germany flared up again. There can be no doubt that the appearance of the Mustangs constituted a major turning-point. They were given greater freedom of action; from now on the fighter escorts no longer stuck close to the bomber formations. They now could seize the initiative to seek out and hunt down their attackers. Once the Mustang formations dominated the airspace, it was inevitable that German fighter losses would rapidly rise.

Allied air forces struck at the armaments industry and at communications targets on a massive scale. Other attacks on airfields in Germany served as preparation for the Allied airborne operation about to be launched at Arnhem.

On 1st January 1945, code named "Operation Baseplate", the Luftwaffe staged its frantic but forlorn attempt to wrest air superiority from the Allies. At dawn that day more than eight hundred German fighters and fighter-bombers attacked Allied airfields throughout Belgium and Holland. Great secrecy had been achieved by the Luftwaffe in the assembly of this force and there is no doubt that many Allied airfields were caught off-guard and considerable damage was done. These losses were a minor setback to the Allied air forces. Most aircraft were destroyed on the ground; therefore, few pilots were lost. Within days most units were fully up to strength again. The Luftwaffe, on the other hand, lost more than two hundred aircraft and pilots.

These pilots were men who would otherwise have sold their lives dearly in the defence of their country; many of them were irreplaceable. The Luftwaffe's folly of New Year's Day was not only forlorn but proved disastrous to the long term Defence of the Reich. Not only were losses to Allied planes, anti aircraft and fighter defences severe, but as the German aircraft returned over heavily defended areas of their own front at low level, they were fired on by 'friendly' flak. The German gunners took the heaviest toll of all. Casualties totalled two hundred and fifty five, including nineteen unit commanders. Total aircraft losses are estimated at least three hundred. Hereafter, little of the Luftwaffe was seen over the Western front during January and February 1945 since most aircraft had been transferred to the East.

There seemed little left to stem the inexorable advances by the Allies from west and east; the Ardennes offensive had crumbled and Baseplate had failed to blunt the Allied tactical strength.

The Luftwaffe now threw its last reserves into the air battle. The controversial Operation Baseplate delivered the German fighter force its final mortal wound and sealed its fate, what happened from then on was no more than a dying flicker.

As the Germans were abandoning their air force bases, the much maligned Luftwaffe prepared to launch their last great offensive. By April 16th, Soviet spearheads were crossing the River Oder and began to establish beachheads. The German formations, outnumbered two to one in manpower and four to one in equipment, fought valiantly. That afternoon, every Luftwaffe unit available for combat operation was thrown into the desperate battle.

The flow of Soviet troops and equipment continued almost unmolested by the air attack. By the early hours of the 18th, the Soviet Army crossed the river. German defense positions crumbled in the face of superior numbers and firepower.

At the same time, inside Adolf Hitler's Berlin bunker, the situation was nearing a climax. Hitler began to issue orders to many depleted Luftwaffe combat units to attack the Soviet formations at once.

Luftwaffe officials tried to talk Hitler out of the idea. The situation deteriorated further when on the morning of April 21st Berlin was shelled by Soviet long range artillery.

Hitler was furious and demanded an explanation regarding the Luftwaffe's absence from the fight. Hitler was unaware that most Luftwaffe units were gone and those that remained operational lacked sufficient fuel and ammunition to mount an effective attack.

The Luftwaffe was encircled in ever smaller pockets and surrounded on all sides. All that was humanly possible to relieve Berlin was being done.

Hitler now came to the conclusion that the war was lost. Many of Hitler's inner circles tried to convince him to leave the capital but he refused. Meanwhile, Göring planned to take over command of the Third Reich. He seized the opportunity and, before all communication was lost with the Berlin bunker sent a cable directly to Hitler, stating he would take over the German government as its official head. Göring even demanded that Hitler respond to the cable before 10:00pm of that day. Hitler lambasted Göring calling him a traitor.

Enraged by what he saw as a stab in the back from a trusted NAZI. The Führer sent out orders to arrest Göring at once, Thus, at the time of its greatest peril, the Luftwaffe was left without leadership.

Meanwhile, the Luftwaffe continued its desperate attacks against the Soviet columns. On the afternoon of the 27th a force of Fw-190 fighters attacked the Soviet's Oder beachhead. As they approached the target area, a massive anti-aircraft barrage greeted them. It was a slaughter.

The Luftwaffe was finally defeated by the Allied air forces, not in the air but rendered impotent on the ground through starvation of fuel. Conditions were rapidly deteriorating. Fuel and ammunition had almost run out. On April 30th, Hitler named Admiral Karl Dönitz his successor and committed suicide.

From the moment Dönitz assumed command of the German state, he worked to end the war as soon as possible. Masses of German ground, naval and air formations began to surrender completely.

Almost all of the Luftwaffe forces surrendered to the Western Allies. The organised defence of Germany was at an end; an inevitable end.

~TWENTY FIVE~
THE BATTLE FOR BERLIN

The "Battle for Berlin" was the last major European engagement in World War Two and perhaps the most controversial; because the result of this battle was of immense proportions; not only for the major powers but for the entire world.

By the spring of 1945, the war had moved beyond the point where all previous wars had ended and German military strategy to all intents and purposes had ceased to exist. The German armed forces which had seemed irresistible for the first years of the war were being smashed by superior strength; they were now outnumbered, outgunned and outmatched on every front.

As the Allied armies pushed into Germany from east and west, it finally became clear to more and more Germans that the Third Reich which was supposed to last for a millennium was nothing more than an illusion. In this, the sixth year of war, Germany had suffered four million casualties; the cream of the nation's manpower. Towards the end of this agonising conflict, devastating thousand bomber raids on German cities had become common place.

Most German civilians had long ceased to believe in the Nazi propaganda promising "miracle weapons" which would dramatically reverse the tide of defeat. There were few left in the German military who were genuinely convinced that Hitler's Reich could still win the war. The Reich which had until only recently occupied most of

Europe and had driven deep into Russia to the very gates of Moscow. The Reich that had gained control of more territory than that of the Holy Roman Empire had now shrunk almost to the suburbs of the capital city.

Hitler's once powerful and superbly trained troops had been reduced to a mere shadow of the elite panzer forces which had stormed into the Soviet Union 1941. The dreadful catastrophes of Stalingrad and Kursk had devoured whole army groups and inflicted unsustainable losses. The decimated and war weary Wehrmacht, fighting desperately to survive and prevent total defeat was near to total collapse. In the four winters of fighting in the east their equipment had been destroyed or left disabled in the vast terrain of the Russian snows; serving as eerie reminders of their dictator's vision which had gone horribly awry.

Hitler's purge after the assassination attempt on his life in July 1944 accelerated the decline of the army's officer corps and brought further irreplaceable losses. By now nearly the entire established army command had either committed suicide, been executed, or were interned in concentration camps. Following the July plot, no one had dared to suggest to the Führer that the Russians might one day threaten Berlin; for fear that this would have been construed as defeatism and a criticism of Hitler's conduct of the war. Therefore, the organisation for the defence of Berlin was an unbelievable mess and any chance of saving the capital had long since vanished.

The coming battle for Berlin, however, could no longer be ignored by Hitler who was by this stage acutely aware that he neither had the manpower nor equipment to accomplish its defence. The remaining strength was not enough to affect the outcome of the battle, but was enough to prolong the agony. For Hitler, isolated under six metres of earth and concrete and out of sight of the destruction rolling in on him there was to be no respite. Now in the last days of his life, Allied victory was no longer in doubt.

This was the situation in Nazi Germany in 1945. So why did the Germans knowing that the cause was irretrievably lost continue the unequal struggle against hopeless odds; and why did they not bring the war to a swift end thereby sparing unnecessary suffering? The answer to this question is that even at this late stage the Germans were still dominated by their dictator Adolf Hitler who demanded, and got, absolute obedience. His will alone was all that counted, and that was the most critical single factor for the continuation of the war. Citing Frederick the Great, he informed his generals that it was the strength and determination of the leadership which decided whether wars were won or lost. While he lived, Germany would go on fighting as he repeatedly said "until five past midnight".

He was convinced that the Soviets, having suffered enormous losses, had overextended themselves and had no operational reserves left. Therefore, the remaining, severely weakened Russian armies would easily be destroyed. Hitler constantly underestimated his enemies and frequently dismissed German intelligence reports of the massive troop concentrations being built up against them. In refusing to accept that the war was as good as lost, his interference in operations became more and more irrational. He repeatedly departed from original plans and these miscalculations would cause enormous logistical problems and fatal delays; all of which led to the great German disaster at Berlin in April 1945.

Guderian, Hitler's foremost panzer general, always claimed that Hitler was never capable of visualising anything larger than a division, which represented the limit of his competence. He was, however, circumspect when he made these comments; anyone who objected to Hitler's meddling was either silenced or ruthlessly eliminated. So, rather than argue with their leader the German generals who came into contact with Hitler chose to fall in line with his insane plans; despite being shocked at the decrepit extent of his physical decline. It is a biological fact that when the territory of an animal is threatened, it will fight to the death, regardless of the odds, and so it is with man.

In 1940, when driven back into their island, the British had good and sound reasons to accept an accommodation with Hitler. However, the British, seemingly illogically in the circumstances of the time, chose to fight on. That same deep animal instinct now took possession of the Germans.

Hitler's somewhat fanciful scheme was to continue the war in the hope that Churchill's grand, if somewhat fragile alliance, consisting of Great Britain, the USA, and the USSR, would ultimately collapse. He reasoned that his enemies would become exhausted, quarrel amongst themselves and eventually agree to a negotiated peace. Surely, he argued, the Western Powers would realise that Bolshevism was their real enemy and join Nazi Germany in the common crusade. When you consider the advent of the cold war shortly after the Second World War ended may suggest that he was not all that far wrong in hoping for this split. Therefore, rejecting the advice of his generals and refusing to leave Berlin, he awaited the dramatic turn of events that would hopefully save his crumbling empire. The Battle of Berlin undoubtedly ranks along with Leningrad, Stalingrad, Hiroshima, and Nagasaki as an example of the horror and destruction accomplished by modern warfare.

Did certain Western Allied military commanders fight their own war, in which personal glory appeared more important than a successful military and political end to the war? Did this indiscretion prevent the launching of the main offensive on Berlin from the West? Is this allegation without foundation? Perhaps not, because later developments were to prove that a critical opportunity was lost, not only to win the war but to win the peace as well. This was the moment of truth for the Allied Supreme Commander General Eisenhower; who's more cautious "broad front" strategy was diametrically opposed by the British and Montgomery in particular.

Unquestionably, it most certainly would have been possible for the Western Allies to seize the initiative and strike a blow for the future of Europe and the world by taking Berlin first. Their failure to do so was an obvious indication of the conflict between British and American commanders; highlighting the seriousness of the disagreement on how the war should be finished and bears significant testimony to the failure of war time coalitions.

Going back to August 1944, Montgomery had been agitating to launch a major offensive into North Germany. In retrospect this may have been logistically impossible. But in early 1945 when the defeat of Germany had become certain and the freight trains from Antwerp were pouring in troops and supplies, it undoubtedly was possible. Unlike Eisenhower, Montgomery realised the political implications of the current military situation and believed that it was vital for the Western Allies to capture Berlin. Therefore, Montgomery once more argued for a single thrust through Germany towards Berlin; but he upset the Supreme Commander by demanding that he should be placed in command of this operation, tactlessly stating publicly that he had "saved the day" for the Americans in the Ardennes. Eisenhower took exception to Montgomery's offensive attitude, informing him that if he was dissatisfied with his role, then the issue would have to be settled by a higher authority. The threat of the sack was abundantly clear; Montgomery would have to knuckle under or depart. It took enormous diplomatic activity behind the scenes, particularly by Churchill, to attempt to soothe over the affair and ease the strains.

Marshal Foch, the Supreme Allied Commander at the end of the First World War, once commented that his experience during the war had caused him to lose much of his veneration for Napoleon; who had always fought against coalition armies. Foch claimed that any reasonable general could easily defeat a coalition. In the Second World War, with a commander in chief other than Adolf Hitler, many German generals would have ruefully agreed.

Like a bank manager Eisenhower reminded the British that they were living on an overdraft and that they could only continue to do so on the bank's terms. Henceforth, the Americans, who were providing most of the logistical backing and with more than twice the number of troops in the field as the British would call the tune. So, regardless of the policy the British urged, the American view was bound to prevail; British strategy was indeed in chains.

The Supreme Commander then proceeded to shock his British Allies by announcing that his objective was no longer a direct thrust to take Berlin; they would continue their advance on a broad front which would facilitate surrounding and isolating German forces in the Ruhr. This would subsequently be acknowledged by friend and foe alike as a major tactical blunder; General Eisenhower seemed unable to comprehend that Berlin had become the strategic centre of Europe. Churchill, on learning of this development, and being fully aware that the Germans in the West were finished; urged Roosevelt to order Eisenhower to make a bold thrust towards Berlin. The

British Prime Minister was horrified at the prospect of leaving the German capital to be taken by the Russians. He knew that if the Red Army conquered Berlin; half of Europe would immediately become communist and there was every possibility that in a few years time the other half would also be gobbled up.

But Churchill's appeal to the dying Roosevelt was in vain; as were British optimistic hopes of an equal voice in the Alliance. Perhaps the rationale behind this is that unlike their British counterparts, most American generals were not trained to consider political objectives as part of military strategy. American tradition schooled their officers never to usurp civilian supremacy; politics were to be left to the politicians. In giving the Red Army free rein to take Berlin, Eisenhower claimed that he was avoiding politics; insisting it was based on "purely military" factors. But, in effect he was doing just the opposite. History was to prove that Eisenhower's misjudgement in allowing the Russians to take Berlin first was one of the most; if not the most significant events of the Second World War. Had the Western Allies continued as jointly agreed, the war would certainly have ended much differently and with very different consequences for Europe and the world.

As it was, the Americans adopted a completely misdirected policy with regard to the Soviet Union; it totally lacked elementary common sense, of which the penalty was still being paid almost half a century later. Returning to the allegations of disunity in the Allied camp; why did Eisenhower deviate from the original plan agreed to by the Joint Chiefs of Staff? One of the reasons for Eisenhower's intransigence is that he had become quite paranoid about a German last stand in a "national redoubt" in the Bavarian Alps. It is now well known that this scenario existed only in the Supreme Commander's imagination; but while it persisted it shaped allied tactics, to the extent that after the enemy forces were trapped in the Ruhr pocket. Eisenhower astonished almost everyone by issuing orders for major Allied forces to be swung south away from Berlin towards Munich to counter this imaginary threat. "Thus making it certain that Berlin would be taken by the Russians"!

This was a grievous political and military mistake, particularly when British intelligence had informed Eisenhower that there was no evidence to support the rumours regarding the so called national redoubt. All that remained between the Western Allies and Berlin was Eisenhower himself; moreover, it had become apparent that he was heavily influenced by his senior commander Bradley, who was still smarting from his embarrassment in the Ardennes and sought to redeem himself. Is it reasonable to suggest that Bradley's advice to Eisenhower was motivated by personal considerations? Bradley realised that if the plan to take Berlin was executed, due to the dispositions of the Allied armies the main thrust would have to be commanded by Montgomery and would include many American troops; this would mean that the kudos of the last victory would go to Montgomery, a man whom Bradley had learned to detest intensely.

It is, therefore, not surprising that it was Bradley who presented the report to Eisenhower which would change the whole course of the final strategy. This document indicated that the enemy was indeed preparing for a last stand in the Bavarian Alps, and it was this report that finally persuaded Eisenhower to make the fateful Berlin decision. Bradley convinced Eisenhower that national prestige was at stake and not personal jealousies; he also informed Eisenhower that this situation was quite unacceptable and would not only offend generals Marshall, and Patton, but would not be tolerated by the American people. He was not prepared to conduct a supporting role to Montgomery and threatened to resign if Montgomery was placed in command; he would ensure that Patton did the same. If this does not constitute military blackmail, then what does?

To alienate the British further, prompted by Bradley, Eisenhower without even notifying the combined chiefs of staff or the Allied governments, proceeded to send a telegram to Stalin informing him that he would not be advancing towards Berlin and would halt at the River Elbe. In the eyes of the British, and it would appear they were somewhat justified; Eisenhower had violated their agreement and usurped the authority of the Allied governments. Churchill and the British chiefs of staff were incensed at this unilateral abandonment of the jointly agreed plans. They also resented the high handed way which Eisenhower had exceeded his authority by communicating directly with Stalin over their heads. This unprecedented step was to have far reaching political implications for post war Europe, and would sow the seeds for a conflict potentially greater than the one coming to an end.

For the first time in almost three years of close cooperation, the British Prime Minister was furious with Eisenhower; stating the Supreme Commander was hell bent on confusing an already confused situation, and he was now more than ever convinced that Eisenhower was a second rate general. Churchill believed that time was running out and the war would be lost politically unless the Anglo-American forces reached Berlin before the Russians. He expressed that the Americans appeared to be incapable of accepting what was obvious." It was the Russians who were now the mortal danger".

Churchill's suspicions about Stalin's post war aims had grown steadily since Yalta; arguing that the Russians had already reneged on their promises to hold free elections in Eastern Europe, which the Soviets were slowly swallowing up. Consequently, the Western Allies were entitled to advance as far east as possible and stay there, regardless of the Yalta treaty. The tragedy of the last weeks of the war was that the political destinies of the peoples of Europe lay in the hands of a few American generals. The dying Roosevelt, according to the British Prime Minister had abdicated his responsibility to Europe and the free world by passing full powers of decision making to General Marshall, his Chief of Staff. Marshall was hopelessly out of his depth on European affairs; he also regarded the British to be a spent force as a world power and was more interested in seeking Russian assistance against the Japanese in the Pacific.

Marshall warned Roosevelt that casualties would be astronomical unless Russia joined the battle to defeat Japan. He expressed the opinion that it was more important to conciliate the Russians than to remain on good terms with the British. Therefore, encouraged by Marshall, and manipulated by Bradley, Eisenhower placated the Soviets, by disclosing to them his plans. He also held back Patton from driving into Czechoslovakia, which could have been achieved with ease. Quite apart from the immense prestige the Russians stood to win by capturing Berlin, there was the additional danger that they would reach the North German ports first and thus gain an outlet to the North Sea and the Atlantic. They would thus be in a position to threaten vital Western interests and the post war balance of power in Europe. British warnings of the danger were ignored.

In the history of war from Alexander the Great to Mao Tse Tung, the world has not produced more than a handful of men of genius capable of commanding armies. The name of Napoleon springs to mind, perhaps Wellington or Marlborough; such men, however are not thrown up generation after generation. It is certain that nothing even approaching genius touched the American commanders who served the Allied cause in Western Europe in 1945. Perhaps of them all, only General Patton was truly capable of commanding an army, and it would be a travesty to suggest that General Bradley's right hand knew much about what his left hand was doing.

When the Allied armies eventually reached the Elbe on 11th April, the plea was again put forward by the British that the Elbe should be crossed in force and a determined drive be made on Berlin; but once more Eisenhower was adamant and refused. He was again encouraged by Bradley who advised that the capture of Berlin would cost one hundred thousand Allied casualties. This was at a time when the Germans in the West had been decisively beaten. Instead of attacking across the Elbe towards Berlin with Montgomery's 21st army group, which had been specially prepared for this task; Eisenhower, held back to permit Bradley the leading role, relegating Montgomery to guarding Bradley's left flank. When they breached the Elbe defences, the Allies experienced minimal opposition; it was to all intents and purposes no more than a mopping up operation. By the speed of their advance and low casualties they showed that Bradley's estimate of one hundred thousand casualties to be ludicrous.

Bradley, with Eisenhower's approval, actually ordered his troops back across the Elbe when Berlin, almost undefended from the West, was only sixty miles away. There they waited to the astonishment of the Germans who could not understand why the Anglo- Americans did not make a dash to Berlin; later admitting that they had virtually nothing left in the West with which to stop them. Hitler's last desperate order had been to turn the whole Western front about face and bring the German troops into battle with the Russians. This was intended to show the Western Powers that

Germany's real battle was against Bolshevism. When later questioned, some of the German troops stated that it was their belief that the Americans were coming to their aid in a new war against the Soviets that sustained them.

One last desperate appeal was made to convince Roosevelt of the vital importance of capturing Berlin. In a cable to the American President, Churchill stated that British intelligence had information that the vast majority of the population, due to their terror of the Red Army, desperately wanted the Anglo- Americans to take Berlin. From the British viewpoint, they were completely exasperated, realising that Eisenhower had made a colossal blunder. Churchill stated that the Allied commander had his armies in the wrong place and heading in the wrong direction. However, if the British were unhappy with Eisenhower, America's other main ally in this war was more than content. The political and military leaders of the Soviet Union were ecstatic with Eisenhower's assistance in establishing a communist empire in Eastern Europe.

Of significance leading up to the battle for Berlin in 1945 was that Hitler had not died in the July plot on his life, but Germany's last chance of finding a rational way out of the war had. By the autumn of 1944, Germany's foes on the Eastern and Western fronts, after their successful summer offensives had run out of steam. A more prudent German government would have used this pause to seek an armistice and escape the inevitable final wave of destruction. But Hitler refused to acknowledge that Germany was militarily bankrupt, and Nazi propaganda now steeled the people of the Third Reich to resolve to fight to the bitter end. As a result, hundreds of thousands more were to die needlessly. Many could see no other course open to them as effective Nazi propaganda emphasised what the horrors of life would be under occupation. One saying in Germany at the time was "enjoy the war because peace will be dreadful". Terror now reigned; German troops were being hanged and shot in the street because of their unwillingness to die for their Führer.

Advancing deeper into German territory, the Anglo-American armies were opposed by under-equipped and under-trained German troops mostly recruited from the Hitler youth or home guard. In any case, these were essentially the death throes of the weary German people who were battered and demoralised in this pointless sacrifice. Awaiting the final assault, most only wanted this senseless war to come to an end. Fearing Nazi death squads, they were content to surrender after little more than a token display of resistance.

When the Red Army reached the outskirts of the city the stage was set for what was to become the final contest of the war in Europe. For the Berliners, no other shock would equal the sudden appearance of the Russians; this was their darkest period of the war. Hitler issued orders that what was in name only "Fortress Berlin" was to be defended against the Bolsheviks, "to the last man and to the last bullet". This battle, he claimed would decide the outcome of the war. The defence consisted of an outer belt of minefields and anti tank ditches with the underground railway and sewer network also to be used for defending the city.

Meantime, the Soviets were highly pleased to learn that the Anglo-Americans would not interfere, Bradley's troops were turning away from Berlin; heading back to their starting positions across the Elbe. The Soviet troops, many of whom had seen their own towns and villages obliterated by the Germans, were bent on extracting the full price. They had fought their way from Stalingrad across half a continent to seek this moment of revenge. Political commissars were positioned amongst the troops, reminding them of the crimes the Germans had committed against Russian women and children, and the indignities suffered on Russian soil at German hands. They also provided statistics of German looting and destruction in the Soviet Union; the object was to give each man the feeling that he had a personal score to settle. To the Red Army, Berlin was the final objective, the true goal. The fall of Berlin was equated with the destruction of the Nazi system, and the end of the war.

Russian soldiers were motivated to be first into Berlin to demonstrate to the world the supremacy of the communist system in achieving the destruction of fascism. Five thousand tanks and two thousand guns were concentrated in readiness for the battle for the capital. The Red Army, determined and vengeful, launched a massive pincer assault on Berlin; their commander Zhukov did not care how many lives he sacrificed as long as he won in the end. Preceded by a massive artillery barrage that opened up a huge breach, wave after wave of Soviet storm troops threw themselves at the poorly prepared German defences, which collapsed almost immediately and were wiped out. It was evident that the city was ill prepared to withstand an attack of this magnitude. The Germans could not stem the tide and the Soviets swarmed in everywhere, pushing the German lines further back and capturing most of the airfields. Although the fighting remained intense, the end was in sight at Berlin.

The Russians energetically exploited their success; launching further attacks with devastating effect. Russian vengeance on German civilians was swift, merciless, and more often than not, brutal. They exacted a fearful toll of rape, arson, pillage, and wanton murder. The wheel of retribution had turned full circle. The overwhelming superiority of the Soviets enabled them to break through north and south of Berlin to complete the encirclement. Using rockets and artillery, the Russians evened the scores for Leningrad and Stalingrad. Final German resistance crumbled. In Berlin the conditions of the stricken and demoralised defenders were horrific and rapidly deteriorating. The starving population and exhausted troops were reduced to scavenging and eating horseflesh.

Berlin was no Stalingrad; it might hold out against the Soviets through fanaticism and terror for a few days, but no more. It was fast becoming clear that "Fortress Berlin" was a myth; it had never come into existence. Apart from the aged home guard with about five rounds per rifle, almost nothing of the once mighty Third Reich remained. Desperate orders flashed out from the deranged German leader as the remains of the Reich were dissected by the invaders. Confusion led to chaos, order led to counter order and finally everything led to disorder. The German command system disintegrated and the disillusioned German army now ceased to exist as a coherent force.

The German armies had been cut to pieces and defeated in the field, their own field. However, the citizens were too weary to be hurt by the degradation. Berlin was a city with nothing more to lose; three out of four buildings had been destroyed. The civilians could do nothing but lock themselves in houses and cellars, waiting in terror for the advancing Russians. There was no food the water was foul and of course no gas electricity or transport. On 2nd May the remains of the Berlin army surrendered and the battle for Berlin was over. The Soviet red flag fluttered over the ruins of the Reichstag; the dreaded enemy now occupied the capital. The procession of prisoners of war wound its way out of Berlin into captivity in another world. The Russian reign of terror was about to begin; fighting troops were withdrawn to be replaced with an ill disciplined and filthy rabble of Mongolians.

Hitler, on learning of the fate of his fascist partner Mussolini and dejected by the betrayal of Göring and Himmler, committed suicide in his bunker; it was a long overdue event. His body was soaked in petrol and set alight; perhaps this was a fitting end to the man who had plunged the world into flames. His appointed successor Admiral Dönitz, vainly tried to negotiate a separate surrender with the Western Allies; desperately wanting to continue the fight against the Soviets in an effort to rescuing as many German civilians as possible. But the Allies were unsympathetic; refusing to accept any compromise and demanded unconditional surrender.

With their cities ruined and reduced to rubble, the stench of death everywhere and millions of people helpless and hungry; the Third Reich capitulated. The Germans finally accepted total defeat and signed the unconditional surrender documents. This merciless conflict which had lasted almost six years ultimately came to an end on May 8th 1945. The victorious Allied powers did not recognise the successor government of Dönitz and had all of them arrested. Germany would be occupied, not partially as after the First World War but completely. Hitler had always said that Nazi Germany would never capitulate as in 1918. He was right, but in a way which not even he had been able to conceive; the Germany of 1918 had retained a government of its own, and though small, an army of its own. But the Germany of 1945 retained neither

For the victors the war was over, but for the losers the agony would continue; a totalitarian state suffered a totalitarian defeat. The entire armed forces, officers and men, became prisoners of war, and sovereignty in Germany passed from German hands into zones of the four occupying powers; each zone being a mirror of the nation which occupied it. In the assignment of the zones, the British had secured the industry, the Russians the agriculture, and the Americans the scenery.

Berlin was the climax of Hitler's ambition and highlighted the arrogant folly of launching operation Barbarossa; he had gravely under estimated the capability of Stalin's Soviet Union which could sustain massive losses. Hitler's overconfidence and miscalculations caused the German army to find itself in a most perilous situation, and he had no one to blame but himself for Germany's greatest defeat.

No nation has ever been reviled as was the Germany of Adolf Hitler. When the mighty German military machine finally crumbled, seventy million bewildered and terrified people were left friendless and outcast. At the end of six years of bloodletting Europe lay prostrate; many blame the Allied policy of "unconditional surrender" as the principal reason why the war did not end sooner. However, this is not necessarily true. Hitler must bear the whole weight of responsibility for the beginning and the end of this war. He and he alone brought down not only Berlin but the entire Third Reich in chaos, and the end result would be the partition of Europe Had the Germans been able to fight on until August 1945, the outcome would have been the same. The terrifying power of the atom bomb would have forced them to surrender, just as it forced the capitulation of the even more stubborn Japanese.

Thus was concluded the second and last multi-front war which Germany had waged in the twentieth century, and with its conclusion the German general staff which had twice been confronted with carrying out tasks way beyond its strength, ceased to exist. The German military had won many great battles and its generals were acknowledged as the finest tactical and fighting soldiers in modern history.

One can only speculate what might have happened if Eisenhower had seized Berlin ahead of the Russians. The results of battles which are never actually fought must always be a matter of conjecture. There are, however, strong reasons for believing that, if the main Western Allied effort had been directed on Berlin under Montgomery, he would certainly have got there before the Russians;

consequently the political map of Central Europe would have been much different from what it was until 1990.

The question of logistics adds a further dimension to the great debate; these were, however, of Eisenhower's own making, for he had allowed his armies to be sprawled out in a "broad front" strategy which dissipated his strength. More than forty divisions had crossed the Rhine by 1st April, but only eight were positioned on the direct route to Berlin. Most were engaged in the pointless reduction of the Ruhr pocket or chasing after a hollow victory against the mythical Alpine redoubt. German military opinion later confirmed this viewpoint. Eisenhower subsequently made many feeble attempts to vindicate halting his drive at the river Elbe. However, was Truman as the leader ultimately responsible, right in endorsing this action? He was, after all the president who coined the phrase "the buck stops here".

The fault, and fault there most certainly was, as borne by the next generation, can be traced back to Truman's predecessor, President Roosevelt who trusted Stalin's good intentions, almost to the point of absurdity. He maintained that he could "handle Uncle Joe", and establish a partnership with the communists. The clanging descent of the Iron Curtain put an abrupt end to this concept; brought about the Cold War and terminated the brief elation at the defeat of the Third Reich. "The ally of one war becomes the enemy in the next".

Stalin took just six weeks to violate the Yalta agreement by installing communist puppet regimes in Eastern Europe. The single most important factor to arise from this war was that the frontier of the Soviet Union was no longer to the east of Poland but had moved so far west that it was less than one day's drive from the Rhine. Rather than a liberated Eastern Europe, with its people freed from Nazi tyranny; due to Stalin, being taken at face value by Roosevelt, there emerged an Eastern Europe burdened by the yoke of a not dissimilar tyranny. Furthermore, much to the consternation of the Western European leaders; the Americans did not even consider it necessary to negotiate an access corridor from the West Germany to West Berlin. In 1945 the Americans utterly failed to grasp the Clausewitz doctrine that "war is nothing more than the continuation of politics by other means". The Soviets, on the other hand, were clearly focused and never lost sight of their main objectives.

Acknowledging that the American decision not to take Berlin before the Soviets was a fait accompli, Churchill bitterly proclaimed; "there is only one thing worse than fighting with allies and that is fighting without them".

~TWENTY SIX~
THE DEATH OF HITLER

What is the legacy of Adolf Hitler? For no nation has ever been as reviled as that of Nazi Germany.

It is widely believed that Adolf Hitler instigated the largest wholesale slaughter of human beings in the twentieth century and proved capable of changing a well-disciplined civilised nation into one that murdered six million Jews, three million Gypsies and unknown numbers of Russians, Poles, and Germans. How much of that is true will never be known. For history belongs to the victors; and it is the privilege of the victors to tilt the balance of truth in their direction.

Many historians consider Hitler a taboo subject, partly because the examination of his life might be seen as a step towards humanising him and excusing his crimes. Or even to be seen as challenging some of the worst Allied atrocities such as the bombing of Dresden. But it is the job of the historian to examine the facts. And what we know for definite is that Hitler was born on April 20th, 1889. He went to school, he had relatives, and above all he had an insatiable ambition. And probably it was this aspect that set Hitler apart from other men. He was single mindedly relentlessly ambitious and was able to summon his awesome powers of persuasion to motivate the German people for war.

Hitler's own enthusiasm for war began in the First World War. He served in the front line as a headquarters runner, was promoted to corporal, wounded in 1916, gassed in 1918 and decorated for bravery in action; winning the Iron Cross twice. The war had an intense impact on him. He had finally found a purpose and learned much about violence and its application. Following the war Hitler joined the emerging Nazi Party and it was here that he discovered his talent for demagoguery.

He became leader of the Nazis in 1921, was jailed for staging the abortive 1923 - Munich 'beer-hall putsch', and during his prison term dictated a political policy called Mein Kampf, in which he spelt out Germany's need to rearm and strive for economic self-sufficiency. He came to power in January 1933 through democratic means and within months undemocratically destroyed all opposition. He assumed the title of Führer of the German Reich and scrapped the Versailles Treaty.

On the outbreak of the Second World War and against the advice of his military advisors he followed *"intuitive"* tactics that initially gained massive victories. He took direct control of the armed forces in 1941, but after the entry of the United States, the Third Reich was forced into retreat as the tide of war turned against him.

By April 1945 German military strategy to all intents and purposes had ceased to exist. The German armed forces that had seemed irresistible for the first years of the war were now outnumbered, outgunned and outmatched on every front. The dictator's vision had gone horribly awry. He had reached the end of the road. His overconfidence and miscalculations caused the German army to find itself in a most perilous situation, and he had no one to blame but himself for the great German disaster in 1945.

Hitler's interference in military operations had become totally irrational, causing enormous logistical problems and fatal delays. The German generals, despite being shocked at the decrepit extent of their leader's physical decline had little option other than to fall in line with his insane plans, thereby prolonging the agony.

The Allied armies had smashed into Germany from east and west and it was clear to most Germans that the Third Reich, which Hitler boasted would last for a millennium, was nothing more than an illusion. In this sixth year of agonising conflict there were few who genuinely believed that Germany could still win the war. They had suffered four million casualties, the cream of the nation's manpower. And frequent thousand bomber raids were devastating the homeland. The German people were battered and demoralised in this pointless sacrifice and most only wanted this senseless war to come to an end.

They had long ceased to believe in Nazi propaganda promising "miracle weapons" that would dramatically reverse the tide of defeat. The mighty Third Reich that had stretched from the Atlantic to the Caucasus and driven to the very gates of Moscow was now reduced to a narrow corridor little more than a hundred kilometres wide. Hitler's once powerful and superbly trained troops had been reduced to a mere shadow of the elite panzer forces that had stormed across Europe. Catastrophes the like of Stalingrad, Tunis and Normandy had devoured whole army groups and inflicted unsustainable losses.

The Western Allies were at this stage mainly opposed by under equipped and under trained German troops mostly recruited from the Hitler youth or home guard; a pitiful ragtag of middle aged men and adolescents awaiting the inevitable final wave of destruction. Hitler, now in the last days of his life, had survived the 1944 July plot, but Germany's last chance of finding a rational way out of the war had died. Knowing that the cause was irretrievably lost, Hitler continued this unequal struggle against hopeless odds.

He was not prepared to bring the war to a swift end and spare unnecessary suffering. While he lived, Germany would go on fighting as he repeatedly said "until five past midnight". Even at this late stage the Germans were still dominated by Hitler who demanded and got absolute obedience.

Isolated in his bunker, he refused to accept the reality that Allied victory was no longer in any doubt. Hitler issued orders that "Fortress Berlin" was to be defended against the Bolsheviks, "to the last man and to the last bullet". This battle, he claimed would decide the outcome of the war. Nazi death squads were ordered to hang German troops in the street if they were suspected of unwillingness to die for their leader; a leader who at that moment was contemplating the most appropriate method of suicide.

A deranged Hitler, dejected by the betrayal of Göring and Himmler, and learning of the manner of death of his fascist partner Mussolini, decided on his fate. For him there was no other way out and by any yardstick it was a long overdue event. He had no intention of allowing Stalin to exhibit him in a cage and refused to leave Berlin.

In a last deliberate insult to the army whom he blamed for losing the war, he named Admiral Dönitz as his successor. He then proceeded to dictate his last will and testament, stating that he never wanted the war he had unleashed on the world; it had been provoked by the Jews. There was not a word of regret or a suggestion of remorse.

Considering his struggle now to be over, in a bizarre ceremony he married his mistress Eva Braun, who chose to share his fate. Hitler, with moist eyes and shaking with emotion, bid farewell to the motley group assembled in the bunker. Hitler and his new wife retired to their room where he shot himself after Eva Braun had taken poison. It was 3.30 pm on April 30th 1945, ten days after Hitler's 56th birthday. Their bodies were carried to the garden, doused with petrol and cremated.

Perhaps this was a suitably Wagnerian end to the man who had plunged the world into flames. It was broadcast on the radio that Hitler had died fighting at the head of his troops, fulfilling his duty to the end in unity with his soldiers.

The Soviet red flag fluttered over the ruins of the Reichstag as the procession of German prisoners of war wound its way out of Berlin into captivity.

The Third Reich had outlasted its founder by just one week. The colossal performance of the "thousand year" empire that had lasted twelve years was over. The victorious Allied powers refused to recognise the successor government of Dönitz and had them arrested.

Germany was occupied, not partially as after the First World War but completely. Hitler had always said that Nazi Germany would never capitulate as in 1918. He was right, but not in the way that he had conceived. The Germany of 1918 had retained a government and an army of its own. But the Germany of 1945 retained neither.

A totalitarian state suffered a totalitarian defeat. The entire armed forces, officers and men, became prisoners of war, and sovereignty in Germany passed into zones of the four occupying powers. Seventy million bewildered and terrified people were left friendless and outcast and Europe lay prostrate after six years of bloodletting. Many blame the Allied policy of "unconditional surrender" as the principal reason why the war did not end sooner. However, this is not necessarily true. Hitler must bear the entire responsibility for the beginning and the end of this war. He and he alone had brought down the Third Reich in chaos, and the end result was the Soviet occupation of Eastern Europe.

If Hitler had been able to fight on until August 1945, the outcome would have been the same. The awesome power of the atom bomb would have forced the Nazi's to surrender, just as it had forced the stubborn Japanese.

With the conclusion of the second and last war Germany had waged in the twentieth century, the German general staff which had twice been confronted with carrying out tasks way beyond its strength, ceased to exist. The German military had won many great battles and its generals were acknowledged as the finest tactical and fighting soldiers in modern history. But the ambitions of Germany's leaders made victory impossible. The great irony is that Adolf Hitler had formed an army which proved capable of carving an empire surpassing even that of Napoleon and had raised Germany from the humiliation of Versailles, only to plunge it into destruction.

In the bitter years ahead, the German people would look back at this time and wonder if that gigantic dream with its loud mouthed vulgarity, and jack booted cruelty could conceivably have taken place. Hitler's death was anything but a hero's end. In committing suicide he deliberately abandoned his responsibilities and took a way out that he always condemned as the coward's way. Few men in history have gained through mass persuasion and mass destruction, the incredible power of Adolf Hitler whose life and death was quite unique.

In reply to the initial question; his legacy was not unlike that of Shakespeare's Macbeth; he was to prove the victim of his own delusions.

~TWENTY SEVEN~
END OF PACIFIC WAR

The atomic bombings of Hiroshima and Nagasaki brought an end to the greatest war in human history. This event, possibly one of the most contentious of the twentieth century, occurred after the Japanese military leadership, who, in order to save face, chose to ignore President Truman's Potsdam proclamation calling for unconditional surrender. They steadfastly refused to give up and were determined to keep going even if it meant sacrificing twenty million Japanese lives.

This was at a stage when Japan was already effectively beaten and the defeat of Germany had left her alone against the world. The Japanese were defenceless on the seas; their air force was gone; and incendiary bombs were burning out major cities. On top of this, news was being received from every front of Japanese setbacks: in Burma the British Army recaptured Rangoon and on Okinawa the Americans eliminated remaining Japanese forces. And seeking targets of opportunity, allied carrier based aircraft were swarming over Japan.

But, although militarily defeated, it still remained, for defeat to be translated into surrender. For the military obstinately clung to the war that they had started and placed their last hopes on a pitched battle in defence of their homeland. Consequently, in the early hours of August 6th 1945 the Superfortress B 29 bomber Enola Gay piloted by Colonel Paul Tibbets took off from the Marianas and headed towards Hiroshima to deliver her fatal payload of uranium 235 code named "little boy." Under the impetus of war, the most brilliant brains of the scientific world had forged a weapon equivalent to over twenty thousand tons of TNT and capable of generating temperatures in excess of three thousand degrees centigrade.

When the Enola Gay detonated "little boy" over Hiroshima; more than eighty thousand people were suddenly and comprehensively burned to death and a further sixty thousand injured. In the smouldering ashes of Hiroshima, strategic air power reached a terrifying new level. And decades later, Japanese were still dying from radiation

There was little debate at the time regarding the use of the bomb, mainly because the Allies had spent two billion dollars developing it. This is equivalent to twenty six billion in today's terms and represents about half the current US annual trade deficit. However, since the end of the war, the world has debated not only the choice of targets, but also whether the bombs had to be dropped at all.

It is argued that Japan was on the verge of collapse, and with the planned invasion of the mainland still three months away; the Allies could afford to wait. In recent years, a global outcry has arisen that the United States has no moral standing on the issue of weapons of mass destruction; as they are the only nation, ever to use nuclear weapons. There is a consensus that the Americans have blood on their hands and revisionist historians challenge the traditional view that; "dropping atom bombs on Hiroshima and three days later, Nagasaki was necessary to bring the Pacific War to an end, thus saving lives".

It is claimed that air bombardment, sea blockade, and Russian intervention would have forced Japan to capitulate, regardless of atomic bombing. This argument is based on the grounds that the fire raid on Tokyo in March 1945 did more damage than the atom bombs. The Tokyo raid burned out 40% of the city leaving one hundred thousand civilian casualties, one million homeless and half the capital's population displaced.

But, the counter argument is that, the Japanese gave no sign of capitulation after this fire raid; their rigid code of honour, as well as face saving, together with concern for the future of their emperor whom they venerated as a god prevented this. However, memoirs and records now show that the emperor wanted to end the war, but he could not prevail over the military leaders. As more information becomes available, the debate intensifies over whether the atomic bombings were really necessary to save lives. It is accepted that there will always be doves and hawks that try to balance between reducing the suffering of the people in a lost cause and national pride.

For instance, Atomic bombing proponents claim that some Japanese leaders were eager for national suicide. This sentiment is supported by kamikaze attacks, and battles to the death on Iwo Jima and Okinawa.

Paradoxically, while western historians undermine the American moral position, Japanese historians actually bolster it; claiming that the atomic bombings broke the stalemate and proved salvation to the Japanese peace faction, who seized upon the atomic bombing as a new development to facilitate surrender. This was boosted when the second atom bomb was dropped on Nagasaki to provide the doves with a scientific reason to accept the inevitable. The peace faction asserts that without the atomic bombings Japan would have continued fighting. Thereby, necessitating the invasion of mainland Japan planned for November 1945.

When the losses at Iwo Jima and Okinawa are taken into consideration, the human toll of invading Japan would have run into millions. For example, the fight for the five-mile-long island of Iwo Jima, lying seven hundred and fifty miles southeast of Tokyo, lasted five weeks, during February and March 1945. This battle cost twenty thousand Japanese and six thousand American dead.

The most costly and terrible action in the Pacific took place at Okinawa, where the US navy assembled a fleet of one thousand five hundred vessels to get troops ashore. It was a floating city replete with repair shops, hospitals, kitchens, laundries, arsenals of munitions, living quarters, and combat control centres. This fleet, operating in hostile waters six thousand miles from the United States, also provided aircraft carriers for hundreds of aircraft. Okinawa witnessed the greatest Kamikaze raids of the war, and the results were staggering. The kamikaze's claimed forty US ships sunk or damaged beyond repair and three hundred and seventy damaged.

U.S. casualties at Okinawa were the largest of the Pacific war. Over twelve thousand American servicemen were killed and thirty six thousand wounded. Almost 50% of the American dead were naval personnel lost to Kamikaze attacks. Therefore, with the Okinawa experience fresh in their minds, military strategists feared that the invasion of Japan would produce a veritable bloodbath. Especially, when tested on the battlefield, the Japanese will to resist defied all possible estimates. Expected Allied casualties ranged from half a million to one and a half million. Japanese losses would most certainly have been significantly higher, perhaps even obliteration.

The fact that the Pacific War lasted so long is undeniably due to the fanatical commitment of the Japanese soldier. Compared to the five million German soldiers who surrendered to the Allies in Europe, less than 5% of Japanese forces in the Pacific surrendered. It was considered a disgrace to country and family, and a fight to the death was preferred.

But, returning to the bombing debate, there is also the contention that the Americans could have exercised restraint and demonstrated the power of the atom bomb on pre-designated and uninhabited areas.

President Truman rejected this suggestion outright. He decided to use the bomb without prior warning for three reasons;

1. To prevent the Japanese moving allied prisoners of war into these areas.
2. The Japanese could also have shot down the bomber,
3. There was no guarantee that the bomb would detonate.

The President stated; "If the object is to use the bomb to end the war then there is no acceptable alternative other than direct military use".

This may be justified by the Japanese military that still ferociously resisted surrender. Even after two atomic bombings on major cities, after Soviet entry into the war, and when additional atomic bombs were anticipated.

This was due to American intelligence leaking to the Japanese that they had one hundred atom bombs and planned the total and prompt destruction of Japan. In reality, only three more bombs could be produced by September, and seven by December. However, even with this information, the hawks still adamantly opposed surrender, claiming the Japanese would keep fighting to the bitter end for the Emperor in the faith that they shall find martyrdom and eternal life.

But, the emperor accepted the futility of prolonging the conflict and intervened in support of the peace faction. On august 15th he broadcast to the nation that the war must be brought to an end. The Japanese people he stated had endured the unendurable and suffered the insufferable. He announced Japan's acceptance of Allied terms and ordered his forces to lay down their arms.

Many thousands refused to accept what they considered to be a national disgrace, and in symbolic futility committed suicide. Some fanatical diehard Imperial Guards attempted a coup, but it was unsuccessful. Loyal army commanders obeyed the Emperor and refused to join the rebellion. World War Two was over

With a view to assessing the arguments in favour and against the use of atomic bombing we will have to go back to the outbreak of war between America and Japan.

The war started in the Far East when a superbly trained Japanese task force steamed undetected across five thousand kilometres of Pacific Ocean. On reaching their operational zone, on December 7th 1941, just off Pearl Harbour, over three hundred and fifty war planes took off from six Japanese aircraft carriers to launch a combined operation.

The Japanese, being the pioneers of the large carrier strike force, decimated the American Pacific battle fleet, to inflict the worst ever military defeat on the United States. Following this seaborne blitzkrieg, in the first three months of hostilities the Japanese netted conquests the size of the North American continent.
To try and put this into perspective, the Japanese effectively coordinated operations through six time zones and reached within striking distance of India and Australia. In a wave of conquest that is unmatched in both world wars, simultaneous operations were launched against British, Dutch and American possessions; delivering fatal blows to European imperialism. Sheer audacity combined with remarkable duplicity, placed Japan in a position of overwhelming strength throughout the Far East.

Why did Japan attack the United States?

There are several reasons but principally it was because of a tense relationship that had developed between Japan and the United States. This was due to the formers aggressive and expansionist policies, resulting in American economic sanctions against Japan that included an oil embargo.

The three critical reasons behind the U.S. sanctions were:

1. Japan's invasion of China in 1937.
2. The Tripartite Alliance formed in 1940 between Japan and the other two Axis dictatorships, namely Nazi Germany and Fascist Italy.
3. Japan's occupation of French Indochina after France had been knocked out of the war.

The sanctions crippled the Japanese economy and caused them to conclude that they either abandon their objective of Far East domination; or, alternatively, in order to gain the mineral rich Asian territories go to war with America and the Western Allies. Japan was at this stage dominated by the military, and the military leaders decided to go to war and seize the Asian territories, which they considered to be theirs by right. They were prepared to challenge any power or even a combination of powers to secure them. Japan was on the march towards what she earnestly believed to be her destiny. That is Asia for the Asians - under Japanese control.

Admiral Yamamoto the commander of Japan's Combined Fleet proposed a sudden and paralysing knockout on the U.S. fleet at Pearl Harbour. He calculated that the United States would be so weakened by this proposed attack that she would be unable to mobilize sufficient strength to go over to the offensive for about two years. By that time, the conquered territory would be fortified and this would undermine American determination to continue the war. The Japanese speculated that the United States, in the face of potentially unacceptable losses would probably compromise, thus allowing Japan to retain a substantial portion of her territorial gains.

Japanese intentions, however, were not in accord with Japan's capabilities. They did not think clearly through the strategic problems associated with war with the United States. Of far greater significance, they failed to appreciate the cyclone of wrath they provoked by the *"sneak attack"* on Pearl Harbour. To every American the Japanese were perceived as the embodiment of cowardly stab-in-back aggressors.

Pearl Harbour ended all isolationist sentiment and served to arouse the people of the United States to such fury that in a war of vengeance, nothing short of unconditional surrender would satisfy them. They were determined to push the war against all odds to final victory. There would be no negotiated peace on this ruinous war embarked on by Japan.

The traumatic shock of Pearl Harbour effectively united one hundred and thirty million Americans as no other event in their history had ever done. The wave of American hatred against Japan was staggering and resulted in a tremendous influx of young volunteers into the U.S. armed forces. The "day of infamy" mobilised all American resources that would eventually break on Japan with colossal force. Furthermore, the Japanese had badly miscalculated. Escaping damage from the attack, and this was to prove critical to the outcome of the war, were the three all-important American aircraft carriers in the Pacific, the Lexington, the

Enterprise and the Saratoga.

In failing to seek out and dispatch these carriers, Japan committed its first and probably the greatest strategical error of the entire Pacific conflict. This was to prove decisive at Midway six months later when four Japanese aircraft carriers were sunk. Out of necessity, the American fleet carriers had become the major capital ships and forced the US to adopt carrier warfare, which in the long run was to give her victory. Far from being a strategic necessity as the Japanese were later to claim, the surprise attack on Pearl Harbour was in effect a short-term tactical masterpiece but a long-term strategic blunder.

Never in military history has an operation proved more fatal to the aggressor. Once the economic and industrial resources of the US were brought to bear, Japan could never win the Pacific war. For instance, at no time did Japan ever achieve even 10% of US production. The Japanese had taken on the world's most powerful nation. A nation that possessed overwhelming industrial might, vast food producing capacity, huge manpower and was free from bombing. As a result, Japan would pay the ultimate price.

By November 1943 it was evident that the balance had clearly shifted. Despite early Japanese victories, Yamamota's pessimistic forecast that the war would go badly if it lasted more than six months had become true. Japanese gains were immense, but behind the conquests there was neither the manpower nor the industrial resources needed to ensure victory. Half of the Japanese fighter planes for example were still the same design as at Pearl Harbour. Superior technology enabled the Americans to undertake a dynamic two pronged Pacific offensive, supported by operations in China, South East Asia, and the north Pacific.

The Pacific War was the largest naval conflict in history. The scale was astounding and the distances immense. Across the huge expanses of the Pacific, in this new kind of war, two of the most powerful navies in the world found themselves locked in a death struggle.

Every conceivable type of naval activity was represented: carrier aviation battles, brutal surface engagements, bitterly fought nightfights, and stealthy submarine battles. The war was also fought in every possible climate, from Arctic conditions in the Aleutians, to the appalling heat and swelter of the South Pacific.

There were no titanic clashes of armies as experienced in Russia, France or even the Western Desert. Sea power was the instrument of victory and that power was centered on the aircraft carriers to give every task force a most potent weapon in either offence or defence. The carrier-borne planes decimated enemy aerial and naval resistance, allowing the transports to take the fighting men to operational areas. But the ships alone could not take or hold a piece of land. The very idea of warships alone being able to cow the enemy into submission had died spectacularly at the Dardanelles.

When the fleet had done its job, it was up to the footsloggers to go ashore and take the land. In the Pacific war this was a particularly bloody affair. Putting troops ashore in the face of a determined enemy had always been one of war's most dangerous and complicated manoeuvres. For success to be assured, assault forces need both direct air support and effective off shore naval bombardment. Plus special landing craft to bring tanks and artillery ashore. And even then, well dug-in defenders could still inflict a heavy toll.

The Americans gradually gained the upper hand, and in June 1944, their technical and training superiority massacred the Japanese air arm at the great Marianas turkey shoot. In a single day the Americans shattered two fleet carriers, four hundred aircraft and five hundred aircrew. On top of this over one hundred land based aircraft were destroyed. The cost in material was bad enough, but even more serious was the loss of what was left of even partially trained airmen

The clearing of the Marianas introduced the strategic bombing campaign against Japan. B-29 heavy bombers with a range exceeding three thousand miles could now reach most Japanese cities, including Tokyo. The way to Hiroshima was now open in spite of desperate banzai charges and kamikaze attacks. In October 1944, the fate of the Imperial Japanese fleet was sealed at the battle of Leyte Gulf; staking all on victory, capital ships and carriers steamed out to do battle with the largest US armada yet seen.

This mighty three-day battle, the greatest sea battle in history, resulted in the virtual annihilation of the Imperial Japanese Fleet. Of major significance during this battle, Japanese planes swept in low to attack us naval vessels. But instead of attempting a conventional bombing or strafing attack they deliberately crash dived onto ships. In what the Japanese hailed as a triumphant vindication of human sacrifice, the Americans suffered their first kamikaze casualties

Although Allied troops were acquainted with their enemy's sacrificial nature, they were completely unprepared for what came out of the sky during the Battle of Leyte Gulf. Great heroism cannot be doubted, however, the former cold professionalism that had carried the Japanese to the high tide of victory in 1942 was gone. The Japanese saw no other option than kamikaze attacks; their aircraft and aircrews being totally outclassed by their American counterparts. Any aircraft that could get off the ground could be converted into human missiles. They also developed and produced special aircraft solely for "no return" missions.

Rudimentary training bases were set up for seven-day courses on kamikaze tactics. Little training was received, only takeoffs', training on landing was not needed. After holding ceremonial rituals, and writing farewell poems, Kamikaze pilots would fly off to die in planes containing 550-pound bombs. They were often university students, motivated by obligation and gratitude to family and country. Thousands of young eager volunteers came forward; only too willing to show veneration for the emperor by killing themselves on an enemy ship and taking as many as possible of their enemies with them. By the time the atom bombs were dropped; seven thousand five hundred Kamikaze volunteers had been killed, one hundred and twenty US ships had been sunk, and many more damaged. Three thousand allied sailors had been killed and another six thousand wounded.

Their courage was never in doubt but the intelligent use of it was. Ironically, the kamikazes and the sacrificial philosophy behind them were one of the reasons why President Truman decided to drop the atomic bombs. The advantage of employing sacrificial dive-bombers is that they are difficult to shoot down. The fear and panic that Kamikazes generated made ships' gunners so jittery that they fired on their own planes returning from strikes on Japanese targets. The effect of these attacks on American morale was immediate and devastating. Just the anticipation of kamikaze attacks drove some American sailors insane.

For the remainder of the war the allies were subjected to unrelenting kamikaze attacks. Airborne kamikazes were not the only units in action. The Japanese developed other suicide weapons. They had the human torpedo called Kaiten; the rocket propelled human bomb named Okha and the suicide fast sea craft Shinyo.

The damage done by Kamikazes was colossal. The losses would have doubled or tripled if the invasion of the Japanese mainland had gone ahead. Suicide tactics were a vital part of the final Japanese defense.

Probably the greatest kamikaze sacrifice was the giant battleship Yamoto that only had enough fuel for a one-way trip in April 1945 to Okinawa. She was expected to deal out maximum punishment on the Americans before beaching herself and firing off the last of her huge 18-inch shells. However, she was caught and sunk by American aircraft as she left Japanese waters, going down with two thousand four hundred men. This action heralded the end of the dreadnought age. It was the last time that a battleship was sunk by enemy action at sea.

A further significant factor in the Pacific is that the prolongation of the war against Germany adversely affected resources, and ended hopes of the invasion of Japan in early 1945. It was only after Germany's defeat in May 1945 that the Allies were able to turn their full resources against Japan; thus setting the stage for the ultimate act of the Pacific war, the invasion of Japan, the Pacific D-Day. A task that demanded maximum Allied effort, for it was expected that the Japanese would defend their motherland with great tenacity.

The bloody conquests from Guadalcanal to Okinawa had demonstrated to the Allies that Japanese fanaticism increased as their homeland was approached. Determined and suicidal resistance had been encountered all the way across the Pacific that was intensified the nearer they got to the Japanese mainland. At Okinawa twenty five thousand Japanese troops hemmed in by naval and air bombardment had held off American forces six times larger for one hundred days. Every inch of ground, despite the pounding of naval guns and carrier planes, had to be cleared with rifle, grenade and flamethrowers. The bloody assaults on Iwo Jima and Okinawa provided a bitter foretaste of what the Allies could expect when invading Japan.

To counter the invasion, the entire Japanese population was mobilised. Many military commanders viewed this as the trump card for Japan's defense. They believed that Japan retained the capability to win this one final battle and amassed their kamikaze weapons. Theoretically the kamikazes could wipe out the entire Allied fleet expected to invade Japan.

The Japanese people were rallied akin to Churchill's "We will fight them on the beaches etc." and were told if they did not take an enemy soldier with them, they did not deserve to die for the Emperor. Two and a half million Japanese regulars faced six hundred and fifty thousand Allied troops. Eight million local volunteers armed mostly with bamboo spears, were prepared to die for their Emperor. They were ordered to fight to the end. Perhaps all armies expect their troops to do this, but only the Japanese, in the modern era, have done this with any consistency. One might describe this as folly or insanity, but it is still no less a trial for the men trying to destroy such resistance. After three years of crushing defeats, Japan was in a desperate plight, regardless of the assurances of her military leaders that exhorted the war weary and homeless population to even greater efforts in defense of their homeland.

But, Japan no longer possessed the means to wage modern warfare. The Midway and Coral Sea losses were never made good, and by 1945 she had little fuel or aircrews. The bombing attacks had left thirteen million homeless and the naval blockade had severed Japan's life lines to the outside world. Whilst the Japanese could not match the American quality of aircraft, they could in numbers, albeit 50% of which were kamikaze.

The Allies planned to invade the southern Japanese island in November 1945 code named operation Olympic and follow up with operation Coronet, the invasion of the Japanese main island of

Honshu in March 1946. A mighty American armada was assembled for the final blow on Japan, and the stage was set for the greatest mass suicide in history. This was the situation immediately prior to dropping the atom bombs on Hiroshima and Nagasaki that resulted in the Japanese surrender. Thus rendering operations Olympic and Coronet unnecessary.

Japan's surrender ended a dream that had begun with high hopes of victory. The gamble had failed and the empire was now reduced to their shores. To a nation never before defeated in war, headed by an emperor regarded as a god, unconditional surrender was the ultimate disgrace. The bitterness of defeat spread throughout Japan a mood of numbness and exhaustion.

There are three main reasons for Allied victory in the Pacific. The ability of the Americans to keep strong naval task forces at sea for months without returning to base. The leap-frogging offensive to by-pass less strategic outlying Japanese garrisons, and Japan's merchant fleet that was essential for the importation of vital raw materials had been decimated by American submarines.

The destruction of Japanese shipping also prevented reinforcements and re-supply. Japan's submarine strategy was totally misguided whereas the Americans used theirs correctly and reaped the rewards. Aggressive American submarine attacks actually eclipsed the feats of the U-boats in the Atlantic. The Japanese met the first ever conquerors in their history on August 30th, when the first American occupation forces plus a small British contingent arrived at Tokyo bay.

Three days later General Macarthur accepted the surrender on board the US battleship Missouri. The Missouri was selected to represent the navy and the home state of President Truman. General Percival the British OC who surrendered at Singapore and American General Wainwright who surrendered at the Philippines were present at the signing ceremony.

Whatever the moral opposition to the use of the atom bomb, consideration should be given to the armed forces that were spared a violent battle on the Japanese mainland. It must have been a tremendous relief after the ferocious battles they had endured that they would not take part in an invasion of Japan. It is perhaps unseemly to defend the vapourising of two cities that some view as the most monstrous acts of the twentieth century. But it has to be appreciated that the great tragedy of Hiroshima and Nagasaki was not that so many people were instantly incinerated, but that the alternatives were far worse.

Taking into consideration the terrible losses expected from invading Japan, plus the Manhattan Project's cost and the desire to convince the Soviets of the United States' superiority, the only surprise would have been a decision not to use the bomb. President Truman always staunchly defended the atomic bombing. He was convinced that it shortened the war, saved American lives, and avenged Pearl Harbour. To him it was a powerful weapon to be used in a just war against a hated enemy.

In his first public explanation after Hiroshima, he said: "Having made the bomb, we used it against those who attacked us, against those who have starved, beaten and executed American prisoners of war. And against those who have abandoned all pretense of obeying international laws of warfare." The Japanese began this war without warning at Pearl Harbor. They have been repaid many fold and a million lives saved".

He continued;
"Let there be no mistake about it. I regarded the bomb as a military weapon and never had any doubt that it should be dropped on the enemy. In deciding to use this bomb, I wanted to make sure that it would be used as a weapon of war in the manner prescribed by the laws of war".

Most of Truman's advisers supported dropping the bombs, though there were exceptions. Most notably, during the Potsdam conference, Eisenhower indicated that he was opposed to using it because Japan had already lost the war. He told Truman that he hoped the United States would not be the first to use the atom bomb. The problem with Eisenhower's view is that the starting and ending of war is a political act, not a military one. Politicians make war; the military merely obey their political masters.

Taking into account all the money and effort that went into building this super powerful weapon, then only to leave on the shelf as American troops die by the thousands, would, from a political standpoint, have been unthinkable. Perhaps the main reason for dropping the bombs was that America was tired of the war and wanted it over quickly. GI's in the European theatre didn't want to be reassigned to the Pacific; they wanted to go home to their families. Therefore, in all probability, Truman ordered the bombings because delaying the war's end would have been political suicide.

Be that as it may, the cessation of hostilities saw the smoothest and most beneficial military occupation in history. In only five years Japan was transformed into an ally of the west. This was largely due to the architect of American victory in the pacific, General Douglas Macarthur. He embarked on social engineering to refashion Japan along western lines. The Emperor was retained, but he was no longer regarded as a living god.

Macarthur was the most brilliant officer ever to graduate as head of his class at West Point. He was decorated thirteen times for bravery under fire and was the youngest American to be promoted to General and the youngest chief of staff in us history; truly a legend in his own lifetime.

Perhaps no other aspect of World War Two is as controversial as the decision to drop the atom bombs on Japan. But, some advantages did arise from this human tragedy. Shortly after the war ended, the harnessing of nuclear energy, as an efficient power source owes its origins to the Manhattan Project, despite the setbacks of Chernobyl and Three Mile Island.

Mercifully, the suffering inflicted on the peoples of Hiroshima and Nagasaki has never been repeated. However, of far greater significance is the avoidance of conflict between major powers since 1945. This is the longest period in history, and is attributed to the existence of the atomic and later the hydrogen bomb.

The nuclear balance, precarious as it may be, has proved a deterrent. For that, whatever the ethical arguments, the world, and mankind should be grateful.

~TWENTY EIGHT~
PADDY MAYNE; AN SAS LEGEND

Paddy Mayne was a soldier, a solicitor and Ireland and British Lions rugby international. He was also a champion boxer, and a founding member of the SAS. He was one of the most outstanding officers of the Second World War and the raids Mayne led in the North African desert destroyed hundreds of enemy aircraft on the ground.

During the course of the war he became one of the British Army's most highly decorated soldiers, receiving the DSO with three bars; one of only seven British servicemen to be awarded this distinction during World War Two. In recognition of his leadership and personal disregard for danger while in France, where he trained and worked closely with the French Resistance, the post-war French Government awarded him the Legion d'honneur and the Croix de Guerre, the only non-Frenchman to receive such a dual honour.

Paddy Mayne joined the SAS in July 1941 and played a major operational role in the Western Desert. He participated in many night raids deep behind enemy lines where the SAS wrought havoc. Mayne pioneered the use of military Jeeps to conduct surprise hit-and-run raids, particularly on enemy airfields. By the end of the North African campaign it was claimed that he personally had destroyed in the region of one hundred and thirty aircraft. He led the SAS with great distinction through the final campaigns of the war in France, Holland, Belgium, Germany and Norway. Mayne was a fearsome and skillful fighting soldier with the ability to read the situation, anticipate how the enemy would react, and then attack.

Paddy Mayne was born in 1915 in Newtonards, County Down, Northern Ireland. His family was from prominent business people and landowners in the district and he was the second youngest of seven children. Whilst attending Grammar School his talent for rugby became evident. He also played cricket and golf, and showed aptitude as a marksman in the rifle club. On leaving school he studied law at Queen's University, Belfast, and took up boxing, winning the Irish Universities Heavyweight Championship. He lost to the British champion on a split decision.

Mayne earned his first full rugby cap as a lock forward for Ireland in 1937 against Wales. After gaining five more caps for Ireland, Mayne was selected for the 1938 British Lions tour to South Africa; the last of the "Blue Lions". After the war they changed to red. During the Lions tour it is said that he trained by "wrecking hotels and fighting Dockers". He played in seventeen of the tours twenty provincial matches and in all three Tests against the Springboks. The 1938 Springbok team is considered by many to be the greatest of all time. They had just completed their highly successful tour of Australia and New Zealand in 1937, under the captaincy of Philip Nel.

At the time the All Blacks called them the best side ever to visit New Zealand. The Lions; under the captaincy of Sammy Walker, an experienced Irish prop toured without seven of what would have been their first choice players who were unable to get time off from work; an example of the true days of amateur rugby union, when the sport was known as "Rugger."

Paddy Mayne's legal and sporting careers were, however, cut short by the outbreak of World War Two, as it did I'm sure for many young men. After enlisting he underwent training with Queen's University Officer Training Corps., earning a commission in the Royal Artillery. Mayne then volunteered for the newly formed No. 11 Commando unit and as a lieutenant first saw action in June 1941 in an operation against Vichy French Forces in Syria. It was after this particularly brutal and confused operation, in which one hundred and thirty officers and men, a third of the strike force, were wounded or killed, that Mayne's enormous capacity for alcohol and violence got him into serious trouble. Mayne held his Commanding Officer Geoffrey Keyes responsible for the heavy casualties suffered during the Syrian operation.

Keyes who was later to be posthumously awarded the VC in an operation to take out Rommel was considered by Mayne to be arrogant and incompetent. They both argued, Mayne lost his cool and threw a punch to knock out Keyes. He was placed under arrest and waiting trial for court martial when David Stirling arrived on the scene. Mayne's reputation for bravery, which was sometimes characterised as reckless and wild had attracted Stirling's attention. Mayne was well known as an international rugby player and possessed qualities in leadership and courage that set him apart from most men.

Stirling recruited Mayne into the newly formed SAS telling him, "This is one commanding officer you never strike and I want your promise on that". Mayne did and a legendary partnership was born. Although there is no evidence that he was placed under arrest to face the prospect of a court martial, several of his fellow officers allege that this was the first of several Paddy Mayne cover ups.

The idea for the Special Air Service originated with Lieutenant David Stirling, a 6 foot 6 inch Scots Guards officer who had joined the Commando's and was now serving in North Africa. During a parachute jump Stirling's chute got snagged, he crashed heavily and was severely injured. Whilst recuperating in hospital, Stirling began writing down his ideas on strategic raids to penetrate deep behind enemy lines by small, select forces.

Such raiding forces would not require the huge naval backup entailed by full-blown commando assault units. Rather, Stirling proposed using parachute saboteurs to inflict damage on enemy airfields and lines of communications. Men noted for unusual bravery and unorthodox methods would lead the formation to shoot up enemy camps and installations, destroy aircraft on the ground and generally create havoc behind enemy lines before rapidly withdrawing.

Further, Stirling foresaw that the number of men in such a team would ideally be four. Each man in this unit would have at least one special area of expertise. Only the best and bravest type of soldier would be selected for the Special Air Service. When Stirling left hospital he requested a meeting with the deputy Commander in Chief, Middle East General Neil Ritchie. He outlined his proposal and Ritchie immediately saw the benefits. Ritchie recommended it to his commander General Sir Claude Auchinleck. It was approved and Stirling, promoted to Captain was authorised to recruit sixty men. The imposing title of Special Air Service Brigade was bestowed on the unit to confuse the enemy about its true size.

After initial disappointments, it was decided to abandon the idea of parachuting behind enemy lines and to increase mobility, equip the SAS with jeeps, each armed with 4 machine guns. They derailed trains, hijacked trucks, mined roads and destroyed arms and supply depots. Their biggest successes came in December 1941, when in a two week period they wrecked a total of ninety Axis planes. So successful was the SAS that Adolf Hitler himself issued a proclamation stating: "These men are dangerous, they must be hunted down and destroyed at all costs". After this captured SAS men could expect little mercy, most were executed.

Paddy Mayne was soon leading many of the SAS airfield raids. On one occasion, when he ran out of ammunition, he disabled several aircraft by ripping out their control panels with his bare hands. Mayne was in his element, shooting up German aircraft with fast-firing Vickers machine guns. But his favorite activity was driving a jeep through the Officer's Mess after particularly heavy drinking sessions.

The first mass jeep raid took place in July 1942, when eighteen vehicles attacked an enemy air base. First they approached the airfield in single file, then fanned out into line abreast formation and opened fire. A pre-planned flare was fired and the jeeps changed formation again, driving through the airfield in columns of two. They opened fire again, with all machine guns, firing at one thousand two hundred rounds per minute. At least forty aircraft were destroyed and one SAS man killed.

By the end of the North African campaign the SAS had destroyed over four hundred enemy aircraft and vast quantities of war material. The following year the sixty remaining SAS men and six officers formed the nucleus of a fully-fledged regiment the 22nd Special Air Service Regiment. The winged dagger badge and motto 'Who Dares Wins' were created by 27-year-old Stirling.

Stirling was known to the enemy as the 'phantom major'. So, the Germans were jubilant when a unit specially formed to hunt down the SAS captured Stirling now a Lieutenant Colonel near the Mareth Line in January 1943. After four escapes and recapture from Italian prison camps, Stirling spent the rest of the war at Colditz Castle. Meanwhile, his former comrades, led now by Captain Paddy Mayne, continued their audacious raids.

Following Stirling's capture the SAS was reorganised into two separate parts, the Special Raiding Squadron and the Special Boat Section (the forerunner of the Special Boat Service). Promoted to Major, Mayne was appointed to command the Special Raiding Squadron and he led the unit with distinction.

Following victory in North Africa, the SRS sailed for Sicily in July 1943. Mayne's men captured three coastal batteries in the initial landings. In September they went to the "Toe" of Italy together with Commandos to seize the port of Termoli on the Adriatic coast. They landed to the west of the town on 3rd October and linked up with the advancing land forces, but soon the Germans launched a ferocious counter-attack. The British forces suffered serious casualties during several days of heavy fighting. The SRS lost sixty eight men killed and wounded.

This was not the role they were developed for. They are specially trained for hit and run tactics. Mayne's squadron sailed for the UK in December 1943. Here it was decided to form the 1st SAS Regiment with Mayne as Lieutenant Colonel in command. He was determined to return the SAS to its proper strategic role in the coming invasion of France.

A Squadron of SAS was parachuted into France on 21st June 1944, others soon followed to establish bases behind enemy lines.

They operated in cooperation with the local resistance to launch attacks on German lines of communication, such as mining roads, destroying railways and bridges, and ambushing convoys. Mayne was behind enemy lines for most of the next four months and personally led many raids. He had a number of close calls, such as the time he and his squad arrived at their hide-up farmhouse only to find it occupied by Germans. A fierce fire fight broke out before the SAS captured the farmhouse.

By the winter of 1944 most of the SAS units had linked up with the advancing Allied forces.

They had tied down thousands of enemy troops and destroyed huge amounts of equipment, as well as calling in many air strikes.

As the front line stabilised during the winter there was little work for them and they were eventually sent back to the UK for further training and re-grouping. On 7th April 1945 Mayne took B and C Squadrons back to the continent for what was to be the last campaign of the war. They were to operate in support of the 4th Canadian Armoured Division, then fighting in north-west Germany.

Mayne's first task was to force a gap in the German lines. The SAS jeeps, although heavily armed and now also partially armoured, were not really suited to this task and soon the column was held up and sustained a number of casualties. Mayne's response was to drive his jeep to the head of the column and then straight at the enemy with all guns blazing, drawing fire away from the pinned-down troops. He then single-handedly rescued a squadron of his troops, lifting the wounded one by one into his Jeep whilst under heavy and continuous fire.

For this action he was recommended for a VC. Although the VC recommendation was signed by Field Marshal Montgomery, Mayne instead received a fourth DSO for this incredible act of courage and leadership under fire. He was, however, denied the Victoria Cross, which he so richly deserved.

After the German surrender in May 1945 the SAS went to Norway, where they had the task of disarming the German troops. In October the SAS regiments were disbanded and Mayne discharged soon afterward. After a period with the British Antarctic Survey in the Falklands and South Georgia, he returned to Newtonards to work first as a solicitor and then as Secretary to the Law Society of Northern Ireland but never really settled back in civilian life.

It has often been questioned why Mayne was not awarded the Victoria Cross; even King George VI expressed surprise at the omission. The answer almost certainly lies in Mayne's abrasive attitude to some of his superiors, combined with the Army hierarchy's negative view of the unconventional tactics of Special Forces. In spite of the effectiveness of these raiders, the British high command at the time never really appreciated their worth.

Paddy Mayne suffered from a crippling back complaint that had been sustained during his army days. The severity of back pain even prevented him from watching his beloved rugby as a spectator. He was ill at ease with the mundane post-war life among provincial lawyers and became reserved and isolated, rarely talking about his wartime service.

On 13th December 1955, Paddy Mayne aged 40, had been drinking and playing poker in a pub not far from his home in Newtonards. He later left, and went on to a friend's house where he continued drinking. Driving homewards in his Riley sports car at 4am, the car collided with a lorry parked with no lights in the middle of the road just a short distance from his home. He was killed instantly. His death was mourned throughout Northern Ireland and the funeral brought the town of Newtonards to a standstill.

Many legends of his incredible life still exist. These mostly tell of incidents in which, after drinking for several hours, Mayne would challenge every man in the bar to a fight, which he would invariably win. Other accounts describe him as a courageous leader of his men.

Mayne was complex and intelligent, though with a ferocious temper that usually surfaced with alcohol consumption. This tendency was said to have become more marked after the death of his father during the war. Mayne was refused leave to attend the funeral resulting in him embarking on a drinking binge and rampage in central Cairo. Mayne is said to have smashed up half a dozen restaurants and beat up a squad of Redcaps.

There is no doubt that many stories of Mayne's drunken brawls are exaggerated, yet David Stirling himself commented on Mayne's frightening rages which, he said, explained why he was so brilliant in battle.

~TWENTY NINE~
KENNETH CAMPBELL VC

The events of 1940 had left Hitler the undisputed master of Europe and Britain stood alone. She had survived the threat of invasion and the "Battle of Britain" had been won. But this would prove to be a hollow victory if Britain starved to death.

When Kenneth Campbell's last action took place in April 1941, America and Russia had not as yet entered the war and Britain was entirely dependent on her Atlantic life line for vital supplies. She was extremely vulnerable and the "Battle of the Atlantic" was critical to the outcome of the war. The Royal Navy had to disperse her resources throughout the world, not only to contend with the Germans and Italians but also to maintain vigilance on the Japanese in the Far East. The position of the British Empire was indeed most precarious.

On 22nd march 1941 the German battle-cruisers "Scharnhorst" and "Gneisenau" entered the French port of Brest after completing operations in the north Atlantic. French underground agents informed the British of this development. Churchill, being fully aware of the threat posed by these powerful enemy war ships ordered every effort be made to destroy them, regardless of the risks and sacrifices involved.

These thirty two thousand ton battle-cruisers each had nine eleven-inch guns, and were capable of a speed of over thirty one knots.

They could either outrun or outgun any Royal Navy ship at that time. These two awesome capital ships were now berthed in the dry dock for maintenance.

Kenneth Campbell based at St. Eval, Cornwall with coastal command No.22 Squadron was put on immediate standby to prepare a strike. Until this time, RAF's bombing raids on the German capital ships at Brest had proved largely ineffective. However, when an unexploded bomb was located in "Gneisenau's" dry dock, the situation dramatically changed. The "Gneisenau" had to be moved into the outer harbour while the bomb disposal team rendered the bomb harmless. This was to become the first link that led to Kenneth Campbell and his crew's supreme sacrifice.

The re-berthing of the battle-cruiser was photographed by an RAF reconnaissance Spitfire, in what may have been one of the most decisive photographs of the war. The "Gneisenau" having left the security of dry dock, now lay exposed in Brest harbour and presented an immediate opportunity for Coastal Command. It was urgently decided to launch an aerial torpedo attack by 22 Squadron's Beaufort's. A strike was planned for dawn the next day on the most heavily defended port in occupied Europe.

Six aircraft were given the task of attacking the German battle-cruiser, surrounded by hills locating almost a thousand anti-aircraft guns. Any aircraft daring to pierce such a curtain of fire would have only the remotest chance of survival. If any mission could be classified as suicidal, this certainly was one. The plan was to send in three Beaufort's armed with land mines to destroy any torpedo nets, thus paving the way for the other three armed with torpedoes.

But, even on take - off, the mission seemed ill fated. St. Eval airbase was rain soaked and two aircraft carrying land mines became bogged down and were unable to take off. The third land mine carrying Beaufort failed to find Brest and returned to base. Under normal conditions the remaining aircraft might have been recalled. But the battle of the Atlantic was at its height and Churchill's directive forbade this option.

At dawn on 6th April 1941 Flying Officer Kenneth Campbell piloted his torpedo bomber towards the "Gneisenau." He was flying on his twentieth operational sortie and was known for determination and unwavering courage in the face of odds.

After circling the target area waiting in vain for the explosions to signal that the torpedo nets had been destroyed, he decided to launch his attack; possibly assuming that the nets had been destroyed. Sighting the battle-cruiser, he dived down to mast height and flew steadily through the blazing flak crossfire. Campbell would have to make sure his torpedo cleared the stone mole guarding the harbour. On his approach, he lined up his aircraft and aimed to drop the torpedo as he crossed the mole, thus giving it the longest possible run to the enemy ship.

He did not have much time from the moment of sighting the "Gneisenau" to the dropping of the torpedo. The distance was almost too close for his torpedo to run effectively. But, without hesitation he skimmed over the water through the concentrated barrage to drop his torpedo accurately towards the "Gneisenau's" stern. The aircraft and torpedo crossed the mole independently. Every anti-aircraft ship and land gun in the harbour was firing at him.

After releasing the torpedo, Campbell immediately pulled up to make a climbing turn, desperate to clear the surrounding hills and reach the sanctuary of low cloud. He was a steadfast man of resolution, a cool head and undoubted courage. He would have reached safety within moments. However, he was hit by a withering hail of fire. Nothing could have survived such a wall of steel and the doomed aircraft went out of control to crash in flames into the harbour waters. There were no survivors and this gallant crew was never to know the results of their courage and determination.

Flying Officer Kenneth Campbell had done his job, his torpedo, running true struck the "Gneisenau" below the water line and the ship began to list heavily. There was extensive damage and only the efficiency of salvage vessels prevented her from sinking. Had she been at sea, she most certainly would have been sunk!

The bodies of Kenneth Campbell and his crew were draped in flags and a guard of honour was mounted as a mark of respect. The Germans, astonished at the bravery of Campbell and his men later buried them with full military honours at Brest. The following morning "Gneisenau" re-entered Brest dry dock. Inspection confirmed that Campbell's torpedo had blown a twelve metre hole in the starboard to flood its turbines and wreck the starboard propeller. It would require six months to repair; in fact the "Gneisenau" was out of action for nine critical months after suffering further damage from RAF bombs.

An accurate assessment of the attack and its effect was compiled by British intelligence. This was supported by enigma decrypts of German messages and resulted in a recommendation for the posthumous award of the Victoria Cross to Kenneth Campbell.

This deserves a closer examination of the real significance of this operation, particularly in the Atlantic.

During their previous sortie, the "Scharnhorst" and "Gneisenau" in just two months had sunk one hundred and sixteen thousand tons of shipping, and the British aircraft carrier "Glorious". The U-Boats by this time were hunting in wolf packs and sinking in excess of three hundred thousand tons of shipping each month. These losses were irreplaceable as British ship yards were being blitzed on a daily basis, and mounting losses by far eclipsed British ship building and repair capacity.

The Atlantic life line was the prime battleground, and Britain, being fully dependent on her sea lanes for vital supplies, was perilously close to losing the war. Hitler was determined to sever Britain's life line with the avowed intention of finishing Britain quickly by blockade, to enable him to focus on the east.

Britain's already precarious supply position would most certainly have been jeopardised had the "Gneisenau" and "Scharnhorst" been able to break out from Brest to rendezvous with the "Bismarck" and "Prinz Eugen." Both the latter two ships sailed into the Atlantic from German waters in May 1941, the month following Kenneth Campbell's action.

Had Kenneth Campbell not crippled the "Gneisenau" and she together with "Scharnhorst" managed to return to the Atlantic. The highly probable merchant shipping losses, coupled with the loss of several Royal Navy capital ships may have compelled Britain to sue for peace. Hitler firmly believed that the combined destructive power of his surface fleet, U- Boats' and Luftwaffe were capable of starving Britain into submission within two months.

Kenneth Campbell's heroic action, in disabling the "Gneisenau" put paid to this dilemma for Britain. Hitler's dream of joining the two battle-cruisers up with the "Bismarck" and possibly later with the "Tirpitz" to form a lethal battle squadron in the north Atlantic was effectively ended.

It is accepted that one single event does not win a war, particularly a war of the magnitude of World War Two. The consequences, however, of the solo torpedo attack launched by Kenneth Campbell was of major proportions in turning the tide of the "Battle of the Atlantic." Further investigation into Kenneth Campbell's outstanding action reveals that this is the only VC awarded based on intelligence reports. All other VC's have been awarded based on eye witness accounts; also Kenneth Campbell is the only torpedo pilot to be awarded the VC. Therefore, Kenneth Campbell's award, combining supreme valour and sacrifice is quite unique.

The Victoria Cross is by far the world's most coveted medal for bravery and is essentially the most democratic. It does not discriminate between ranks and is the highest and most distinguished military decoration awarded to any service man.

In October 1998 the author decided to visit his and Kenneth Campbell's home town Saltcoats in Ayrshire, Scotland for the first time in ten years. He had the privilege of meeting the now deceased Mr. James Campbell who took great pride in showing the VC medal won at such a terrible price by his brother. He also provided copies of the original documents relating to this remarkable event.

The author visited the local war memorial and was horrified to find that the letters VC were not inscribed next to the name of Kenneth Campbell. It was apparent that Kenneth Campbell VC, one of the country's greatest heroes was not honoured or even acknowledged in the town of his birth. His brave and unique exploit was relatively unknown.

The author decided to launch a campaign from KZN, South Africa to have this situation rectified. The regional authorities in Ayrshire were contacted to inform them of developments and surprisingly the area convener Samuel Taylor remembered the author from school days. Various other interested parties were also contacted and the campaign then really got off the ground to gain momentum.

A dedication ceremony was held on the day after the 59th anniversary of his death on 7th April 2000.

For more than half a century Campbell's singular valour had been known of and recognised by few people outside his own family. That was rectified at a ceremony in the town that paid long-overdue tribute to its forgotten hero.

The author was invited as a guest and flew over to attend the occasion, and en route visited Brest to view the scene of the action and visit Campbell's grave. The ceremony was organized by the Regional Council and brought the town of Saltcoats to a partial standstill. A delegation from No 22 Squadron at North Devon, headed by their Officer Commanding, Wing Commander David Simpson, flew up to Scotland in a Sea King helicopter.

Towards the end of a formal luncheon, an extremely emotional Mr. James Campbell presented the Victoria Cross medal to No 22 Squadron. He stressed that it should be regarded as an award to all the crewmembers of the Beaufort and that the squadron must hold it indefinitely to provide a source of pride and inspiration to all those who serve in No 22 Squadron.

In his reply, an equally emotional Wing Commander Simpson stated that this was the proudest moment in his entire career. Immediately afterwards, everyone moved to the dedication site where a Guard of honour was formed for the Standard Party, bearing the squadron's Battle Honours.

The Queen's representative, Lord Lieutenant, Col John Henderson, arrived. He was accompanied by an RAF colour party and a group of ex-servicemen including Royal British Legion office bearers and a survivor of the raid, Ron Bramley then aged 79. Lord Lieutenant, Col John Henderson inspected the Guard of honour.

Councillor Taylor gave the Opening Address in which he thanked the South African Military History Society for initiating and contributing towards the occasion. This was followed with the reading by Wing Cdr Simpson of the London Gazette 1942 citation announcing that Flying Officer Kenneth Campbell had been awarded the Victoria Cross posthumously.

A brass plaque was unveiled above a cast iron bench near the town's war memorial. The bench displays the profile of a Bristol Beaufort, bearing Campbell's name in gold lettering, with the inscription "For Valour".

Overhead, an RAF flag was flown by a Sea King helicopter of 22 Squadron in a fly-past tribute. Then Rev. Sandy McCallum gave a Prayer of Dedication and Blessing. The ceremony ended with a traditional bagpipe lament, played by Pipe Major Jim Butler.

National and local newspaper interviews followed and a few days later the author participated in a live talk show on BBC radio. Several other prestigious journals and organisations also published articles of the occasion. In all of these publications the South African Military History Society received good coverage. On returning to South Africa, the Natal Mercury published a most interesting article.

The Natal connection on Kenneth Campbell's action has a twist in the tale. As previously stated, the name of the German battlecruiser that Campbell attacked at Brest in 1941 was the "Gneisenau." A South African pilot from Natal named Stuart Simson had previously attacked and damaged this huge ship at Kiel Harbour only a few months previously in August 1940. He piloted his Wellington aircraft through vicious flak to a height of one hundred feet to ensure accuracy and dropped his bomb load to inflict severe damage. The result was that was that the "Gneisenau" could not put to sea for three months at this critical stage of the war.

Pilot Officer Simson was awarded the DFC for this action, survived the war, enjoyed an illustrious career with over twelve thousand flying hours and promoted to the rank of RAF Squadron leader. At the war's end he was awarded the AFC. His son is currently living in Umhlanga Rocks

Returning to Kenneth Campbell's fateful action in April 1941; some month's later Hitler became so concerned for the security of his two battle cruisers that had been joined by the heavy cruiser Prinz Eugen that he ordered them to be brought home from Brest to the safety of German waters.

This led to the "Channel Dash" in February 1942. All three ships made it, and the operation proved a huge propaganda success for Germany and a crushing humiliation for the British. It was a tactical victory for the German Navy was also a strategic gain for the Royal Navy. The Brest Squadron no longer directly threatened the Atlantic convoy routes.

After repairs the Scharnhorst was sent to Norwegian waters to join the Tirpitz in threatening Soviet convoys. Her fate was sealed in December 1943 when she was trapped off the North Cape by Admiral Fraser's Home Fleet. The Scharnhorst was first detected by the cruiser HMS Belfast on which Able Seaman Thomas Brady was serving. The author's late father!

Scharnhorst was put out of action by the heavy guns of "Duke of York" and finished off by the torpedoes of the escorting cruisers and destroyers. Only thirty six of a compliment of almost two thousand survived.

Two weeks after docking in Kiel, the Gneisenau became the target of a huge RAF attack by one hundred and seventy eight bombers and was struck on the foredeck. Contrary to normal practice, ammunition had not been unloaded. This was a fatal error and the resultant fires set off a chain reaction and explosion that devastated the ship and killed one hundred and twelve crewmembers. Gneisenau was towed to Gotenhafen in March 1942 where it was decided to decommission her. B and C turrets were removed from the ship and taken to Norway; B-turret being deployed west of

Bergen, while C-turret was deployed near Trondheim.

Gneisenau ignominiously ended the war as a blockship, sunk in Gotenhafen harbor and filled with concrete to deny the port to the Russians. After the war she was raised by the Poles, broken up, and scrapped.

At the war's end the batteries were taken over by the Norwegians and used in exercises from the late 1940's through to 1953, when the guns were fired for the last time and left in decay until 1990. The Norwegian National defence museum realised what a unique piece of history they had, and in cooperation with local volunteers started a process of restoration to its present condition. The Trondheim gun turret is the only German World War Two vintage naval main gun turret left that has not been either dismantled or is resting on the bottom of the ocean.

www.ingramcontent.com/pod-product-compliance
Lightning Source LLC
LaVergne TN
LVHW050757061126
851857LV00012B/597